VENUS,
Near Neighbor of the Sun

Lothrop Books on Astronomy by
ISAAC ASIMOV

JUPITER, THE LARGEST PLANET
ALPHA CENTAURI, THE NEAREST STAR
MARS, THE RED PLANET
SATURN AND BEYOND
VENUS, NEAR NEIGHBOR OF THE SUN

VENUS,
Near Neighbor of the Sun

ISAAC ASIMOV
Diagrams by Yukio Kondo

Lothrop, Lee & Shepard Books
New York

Printed in the United States of America.

First Edition

1   2   3   4   5   6   7   8   9   10

Library of Congress Cataloging in Publication Data

Asimov, Isaac, (date)
    Venus, near neighbor of the Sun.

    Includes index.
    SUMMARY: A comprehensive discussion of Venus with information on neighboring Mercury, asteroids, and comets. Includes numerous tables and figures.
    1. Venus (Planet)—Juvenile literature. [1. Venus (Planet)   2. Mercury (Planet)
3. Planets, Minor.   4. Comets]   I. Kondo, Yukio.   II. Title.
QB621.A84   523.4'2   80-26700
ISBN 0-688-41976-3
ISBN 0-688-51976-8 (lib. bdg.)

TO ALL THE GOOD GUYS AT THE
DUTCH TREAT CLUB

Contents

List of
Tables

List of Figures

VENUS,
Near Neighbor of the Sun

The planet Venus with its perpetual cloud cover, as photographed by Mariner X at a distance of 720,000 kilometers (450,000 miles) in 1974.

1
Evening Star and Morning Star

HESPEROS AND PHOSPHOROS

After the Sun sets on a clear day and the sky begins to darken, a single star sometimes appears in the west. It will be the only star in the sky for a while and it will grow brighter as the sky grows darker. Other stars will appear, too, as the sky darkens, but not one of them will be as bright—or even nearly as bright—as that first one.

When the sky is completely dark, that first star will shine like a bright diamond, the brightest and most beautiful of all the stars.

It will always be in the west, though; in that section of the sky in which the Sun set and disappeared behind the horizon. What's more, the bright star will be moving toward the western horizon and it will set a few hours after sunset, at most. For the rest of the night it will not be visible.

Because it is only visible in the early part of the night—that is, in the evening—it is called "the evening star."

The ancient Greeks called it that, too, except that they used the Greek language and named it *Hesperos aster* or just *Hesperos* (evening) for short.

Then, too, there are times late at night, not long before the dawn, when a bright star rises in the east. Again, it is brighter than any other star in the sky and shines like a diamond until the Sun rises and blanks out all the stars with its great light. Because this star always appears not long before the morning arrives, it is called "the morning star."

The Greeks called the morning star *Phosphoros*, meaning "light-bringer," because it brought the light of day. When it was seen in the sky, one knew that night was nearly over and the light of morning would soon arrive.

In the Latin language, spoken by the ancient Romans, the evening star is called *Vesper*, which is "evening" in Latin. The morning star is called *Lucifer*, which is "light-bearer" in Latin.

In the Bible, the prophet Isaiah predicts the downfall of the King of Babylon. He refers to the king sarcastically as "the morning star" (or *Helel*, in Hebrew) because that was what the flattering courtiers called him. Isaiah said, "How you have fallen from heaven, O Helel, son of the morning."

When the Bible was translated into Latin, the Latin term *Lucifer* was used for the morning star and the verse in the King James Bible becomes, "How art thou fallen from heaven, O Lucifer, son of the morning." (In more modern translations it is "How you are fallen from heaven, O Day Star, son of Dawn," and "How you have fallen from heaven, bright morning star.")

Some people thought that Isaiah was speaking about certain rebellious angels who revolted against God and were hurled out of heaven and into hell. They gave the name Lucifer to the chief of those rebellious angels, who was thought to be the Devil. For that reason Lucifer, the name of the bright and beautiful morning star, has come to be given to the Devil and is usually taken to be another name for Satan.

ONE PLANET FOR TWO

Most of the stars maintain their same position relative to each other and keep to the same pattern night after night, and century after century. They are "fixed stars."

Some objects in the sky, however, shift position against the stars from night to night. The Greeks called them *planetes* (wanderers) and the word has come down to us as "planets." We know these planets by their Latin names such as Mars, Jupiter, and Saturn.* The Sun and the Moon also shift position against the stars, so the Greeks included them among the planets.

As it happens, both the evening star and the morning star shift position against the starry background and they, too, are planets. Yet you will not find Hesperos and Phosphoros (or Vesper and Lucifer) listed as planets in any astronomy book. There is a puzzle here, and we can find the answer if we watch what happens to the positions of the evening star and the morning star night after night.

Here is one interesting fact you would observe. On those evenings when the evening star appeared as the twilight deepened, there was never that morning star in the sky before dawn. On the other hand, when the morning star appeared before dawn there was never that evening star in the sky after sunset. They existed one at a time, taking turns.

First, there would be an evening star in the sky, appearing very low in the sky just after sunset and setting soon after. From night to night it would appear higher and higher in the sky at sunset, and also brighter and brighter. Finally, it would be about halfway up from the horizon to the zenith (that

---

* If you would like to know more about those particular planets, see my books *Mars, the Red Planet; Jupiter, the Largest Planet;* and *Saturn and Beyond,* all published by Lothrop, Lee & Shepard.

portion of the sky directly overhead) at sunset, and wouldn't set till three hours after sunset.

However, the evening star gets no higher than that. From night to night thereafter, it sinks lower in the sky, getting less and less bright and setting sooner and sooner after sunset. Finally, it is so low in the sky that you can hardly make it out near the horizon in the bright sky after sunset—and then it disappears. You don't see it in the evening any more.

A few days after the evening star disappears, you begin to see the morning star (if you're awake at that time) in the sky before dawn.

At first it appears just before sunrise and can hardly be seen in the bright light of dawn. Almost immediately after the morning star rises, it disappears in the overwhelming light of the Sun.

As you watch, dawn after dawn, however, you will see that the morning star appears at the eastern horizon earlier and earlier and gets higher and higher in the sky before the Sun rises and makes it invisible.

Finally, it rises about three hours before the Sun does and gets about halfway to the zenith before the Sun rises. That, however, is as high as it gets. Thereafter, from night to night it rises later and is lower and lower in the sky by the time of sunrise. Finally, it again rises just before the Sun does, and is blanked out almost at once.

After that there is no longer any morning star in the sky, and there comes a period when it is the evening star's turn again. And all this repeats over and over again.

People who watch this happen repeatedly eventually come to understand that the evening star and the morning star are the same planet; they are not two planets, but one. It is a planet that moves to the east of the Sun and back, then to the west of the Sun and back, over and over again, as shown in Figure 1.

Astronomers among the Sumerian people, who lived in the lower course of the Euphrates River in what is now Iraq,

Figure 1
EVENING STAR AND MORNING STAR

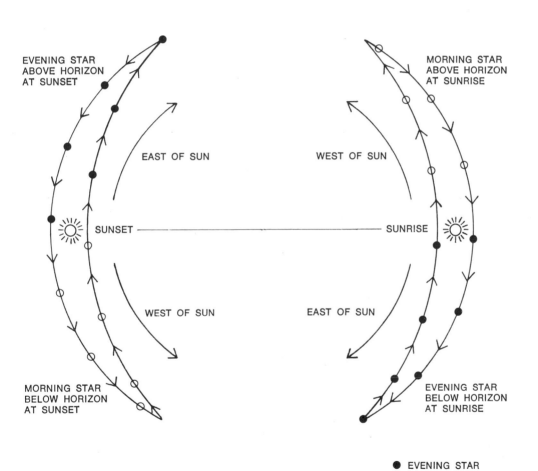

EVENING STAR
ABOVE HORIZON
AT SUNSET

MORNING STAR
ABOVE HORIZON
AT SUNRISE

EAST OF SUN

WEST OF SUN

SUNSET ——————————————————— SUNRISE

WEST OF SUN

EAST OF SUN

MORNING STAR
BELOW HORIZON
AT SUNSET

EVENING STAR
BELOW HORIZON
AT SUNRISE

● EVENING STAR
O MORNING STAR

were probably the first to realize this, perhaps as early as 3100 B.C. They thought that each planet represented a god or goddess, and because the evening/morning star was so bright and beautiful, they naturally named it *Ishtar* after their goddess of love and beauty. They passed their astronomical knowledge on to the Babylonians, who lived in the region later and who made further advances of their own.

The first Greek who realized that Hesperos and Phosphoros were the same planet is supposed to have been Pythagoras of Samos (582 B.C.–507 B.C.). He may have discovered it himself, or he may have learned it from the Babylonians.

In any case, the Greeks borrowed the Babylonian notion of naming the planet for the goddess of love and beauty, and named it Aphrodite, after their own goddess. The Romans took over the notion, too, and using their own goddess, called the planet Venus.

All the names—evening star, morning star, Hesperos, Phosphoros, Vesper, Lucifer, Ishtar, Aphrodite—represent the object in the sky which we call the planet Venus.

All the planets have symbols that were given them in ancient times. The symbol for Venus is something that looks like a little hand-mirror, an object you would expect a beautiful goddess to have about her at all times. The symbol, ♀ , is also now used to represent the female sex.

The ancients knew of seven planets altogether, and the Babylonians set up a seven-day week to correspond. This was eventually adopted the world over, and each day (in western countries at least) was named for one of the planets. The sixth day of the week is *Viernes* in Spanish, *Venerdi* in Italian, and *Vendredi* in French. In all three cases, it means "Venus's day." (In English, however, the sixth day of the week is Friday, and in German it is *Freitag*, being named for Freia, the Norse goddess of love and beauty.)

The ancients knew of seven different metals, and they associated those with the planets, too. The Sun, the brightest of all the objects in the sky, was associated with gold, the most beautiful and precious of the metals. The Moon was naturally associated with silver. Venus, the heavenly object brighter than all others but the Sun and the Moon, was associated with copper, the third most precious metal.

THE BRIGHTNESS OF VENUS

Back about 130 B.C., the Greek astronomer Hipparchus (hih-PAHR-kus) divided the stars of the sky into six groups according to their brightness, which we now call "magnitude." The brightest stars in the sky were grouped together as "first-magnitude," the next brightest as "second-magnitude" and so on. The very dimmest stars, the ones a person with good eyesight can just manage to see on a clear, moonless night, were of the "sixth magnitude."

As time went on, astronomers invented instruments that made it possible to measure the light from stars quite accurately. It then turned out that first-magnitude stars gave off about a hundred times more light than sixth-magnitude stars did.

An English astronomer, Norman Robert Pogson, in 1850 suggested that starting at 6 for the faintest stars, one could move up the numbers by making each magnitude represent 2.512 times as much light as the next larger number. A star of magnitude 5 would deliver 2.512 times as much light as a star of magnitude 6. A star of magnitude 4 would deliver 2.512 times as much light as a star of magnitude 5 and 2.512 × 2.512 or 6.310 times as much light as a star of magnitude 6. A star of magnitude 1 would deliver 2.512 × 2.512 × 2.512 ×

2.512 × 2.512 or 100 times as much light as a star of magnitude 6.

A star known as Pollux, for instance, has a magnitude of about 1 but not quite. Astronomers can now measure Pollux's light so accurately that they can tell it delivers just a tiny bit less light than a star would if it were exactly 1 in magnitude by the standards of brightness that have now been set up. Pollux's magnitude is just a little in the direction of 2. It is, in fact, 1.16.

There are stars that are brighter than Pollux, and their magnitudes are less than 1. The star called Vega has a magnitude of 0.04. It is actually a "zero-magnitude" star, delivering 2.5 times as much light as Pollux does. (However, all stars that have a magnitude under 1.5 are still lumped together as "first-magnitude stars.")

There are four stars that are even brighter than Vega, and the only way of expressing their magnitudes is to go into negative numbers. A magnitude of –1 is 2.512 times as bright as a magnitude of 0; a magnitude of –2 is 2.512 times as bright as –1; and so on.

The brightest of all the stars in the sky is Sirius; it has a magnitude of –1.46. Sirius is 3.7 times as bright as Vega and 11.2 times as bright as Pollux.

Although the ancients thought of a planet as a kind of star (so that we speak of the "evening star" instead of the "evening planet"), the two objects are really altogether different. The stars are large bodies that are so hot they give off light of themselves. Our Sun is an example of a star. The other stars look as dim as they do only because they are enormously far away.

The planets are much smaller bodies than the stars are, their surfaces are cool so they don't glow as stars do, and they don't shine of their own light. The planets we see in the sky are much nearer to us than the stars are (except for our own

Sun), and those planets shine only because they reflect light from the Sun.

The planets are so close to us compared to the stars, however, that the small quantity of light planets reflect can be enough to outshine the largest and hottest of the very distant stars.

Most stars shine with the same brightness, night after night, century after century, but planets tend to undergo regular changes in brightness as they move across the sky. If, however, we take each planet at its very brightest and measure its magnitude, we find that three of the planets are brighter than even the brightest star (see Table 1).

Table 1
THE BRIGHTEST PLANETS

Planet (Star)	Magnitude	Amount of Light Delivered (Sirius = 1)
Venus	−4.22	12.6
Jupiter	−2.55	2.73
Mars	−2.02	1.68
(Sirius)	−1.46	1.00

As you see, Venus, when it is at its brightest, delivers 12.6 times as much light as Sirius does and 4.6 times as much light as Jupiter, the next brightest planet.

The Moon is brighter than Venus. (The ancients considered the Moon a planet because it moved against the background of the stars, but modern astronomers consider it a different class of body, and it is as bright as it is chiefly because it is so near the Earth—much nearer than any other astronomical body.)

The Moon, when it is full and at its brightest, has a magnitude of –12.73, and it then delivers about 2,500 times as much light as Venus does. As for the Sun, its magnitude is –26.91, so it delivers about 470,000 times as much light as the full Moon does, and a little over 1,000,000,000 times as much light as Venus does.

On a clear, moonless night, though, when Venus is quite high in the sky and is at its brightest, it is the most beautiful object we can see with the unaided eye. Unlike the Sun and Moon, it isn't bright enough to wash out other objects. Venus is surrounded by all the stars and planets and is brighter than all of them by far, ruling over them like a queen over her court. In fact, Venus can shine with such brilliance that it actually casts a very faint shadow.

THE MOTION OF VENUS

The motion of Venus—away from the Sun, then toward it, then to the other side, away from it and toward it, then back to the first side, over and over again—made it look as though it were circling around the Sun. If the people on Earth saw the circle edge-on, then Venus would move in the sky just as we see it move.

The first person we know of who suggested this, about 330 B.C., was a Greek astronomer named Heracleides (her-uh-KLY-deez). An astronomer, Aristarchus (ar-is-TAHR-kus), who lived half a century after Heracleides, thought that *all* the planets moved around the Sun. Even Earth did, he said.

Other astronomers of the time would not accept that, however. It seemed to them that the Earth had to be at the center of the universe, solid and motionless, and that all the objects in the sky revolved around it. To account for the fact that Venus seemed to swing back and forth, away from the Sun and back, first on one side and then on the other, they had to imagine a very complicated scheme of motion in which both Venus and the Sun moved about the Earth.

The more closely the old astronomers studied the way in which the planets moved, the more complicated were the explanations they had to devise—so long as they insisted on putting the Earth at the center.

Finally, in 1543, a Polish astronomer, Nicolaus Copernicus (koh-PUR-nih-kus), published a book which explained in great detail the reasons it made more sense to suppose that all the planets, including Earth, moved around the Sun.* For one thing, it explained the motions of the planets in the sky more simply if the Sun, not the Earth, were considered at the center of the planetary system. (With the Sun at the center, the planetary system became the "solar system" from the Latin *sol*, meaning "sun.")

With the Sun at the center, it was easy to explain, for instance, why it was that Venus stayed near the Sun at all times, while Mars, on the other hand, could be at any distance from the Sun.

The explanation rested on the fact that while Venus, Earth, and Mars all moved, or revolved, about the Sun, Venus was closer to the Sun than Earth was, and Mars was farther from the Sun than Earth was.

We can see what this means if we look at Figure 2, which shows the paths, or orbits, that Venus, Earth, and Mars take as they move around the Sun. (We are imagining that we are

---

* The Moon, however, does revolve about the Earth. Because of this, astronomers now do not list it as a planet but as a "satellite" of Earth.

Figure 2
THE MOTIONS OF MARS AND VENUS

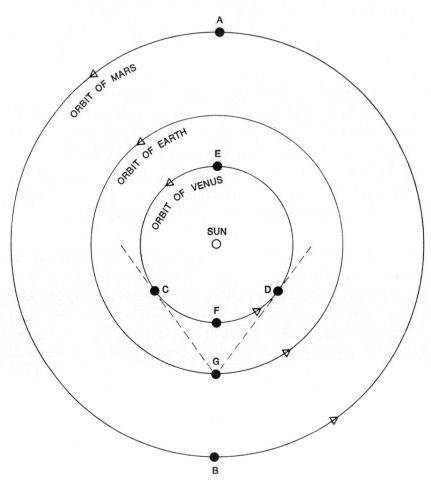

looking at those paths from a position high above Earth's north pole.)

Imagine that Earth is at position G and remains there.*

---

* Earth is actually moving, of course, but that doesn't change the explanation that follows, and it is simpler to understand the explanation if we pretend Earth is not moving in its orbit.

If Mars is at Position A, we can't see it from Earth for it is hidden behind the Sun. As Mars travels onward in its orbit, however, it emerges from behind the Sun and after that it can be seen in the night sky farther and farther from the Sun. By the time Mars reaches position B, it is behind the Earth. We then see Mars at the very opposite point from the Sun. At midnight, when the Sun is shining on the opposite side of the globe from us, Mars is high in the sky.

When Mars is at position A, it is in "conjunction," from Latin words meaning "join together," since Mars and the Sun seem to come together in the sky. When Mars is at position B, it is in "opposition," because it and the Sun are at opposite positions in the sky.

Venus, on the other hand, since it is closer to the Sun than we are, can never be in opposition. It can never get behind us as Mars does.

When Venus is at position E, it is in conjunction. It moves out as Mars does, but instead of sweeping around and getting behind us, it moves to position F, where it is again in conjunction with the Sun. In two different places in Venus's orbit, we find Venus in the same direction as we see the Sun.

In position E it is behind the Sun, and this is the "superior conjunction" (from a Latin word meaning "higher"), because as we look upward at the Sun, Venus is beyond the Sun and so can be considered to be higher than the Sun. In position F, Venus is at "inferior conjunction" (from a Latin word meaning "lower"). Venus is then between us and the Sun and can be considered lower in the sky than the Sun is.

If we imagine Venus at inferior conjunction at point F, it moves along its orbit to the right and gets farther and farther from the Sun (as viewed from here on Earth) until it gets to point D. At point D, it is farthest from the Sun as viewed from Earth. When Venus moves in its orbit past point D, it seems to move closer and closer to the Sun. You can see this if you try to draw dotted lines to other parts of Venus's orbit. Any dotted line you draw to any point in the orbit is nearer the Sun than the dotted line from G to D. Any dotted line you draw outside the one from G to D misses the orbit altogether.

At position D, then, Venus is at "maximum elongation." The line drawn from the Sun to Venus in Earth's sky is then longest.

When Venus passes position D, it continues toward E where it is at superior conjunction, then moves past that to position C where it is at greatest elongation again, but on the other side of the Sun.

THE PHASES OF VENUS

Although the Copernican scheme, in which the planets move about the Sun, explained so much, astronomers were reluctant to accept it because it seemed so odd to think of the huge, solid Earth whizzing through space.

Sixty-six years after the death of Copernicus, however, an Italian scientist, Galileo Galilei, usually known by his first name (gal-ih-LAY-oh), constructed a simple telescope and turned it on the sky. On December 11, 1610, he made an interesting discovery concerning Venus. He was a little nervous about announcing it to the world before he was sure. On the other hand, if he didn't announce it someone else might make the same discovery and get the credit.

Galileo therefore sent a letter to a friend, and in it he said, "*Haec immatura a me iam frustra leguntur o.y.*" This is Latin for "These unripe things are read by me." By "unripe" he meant matters he was not quite ready to discuss. The final "*o.y.*" showed that two letters were left over, and this was a hint that if the letters were all rearranged and the "*o*" and "*y*" were included, there would be a different message.

Once Galileo was satisfied his discovery was correct, he would rearrange the letters and everyone would know how early he had made that discovery. Those who made the discovery later would receive no credit. On the other hand, if the discovery turned out to be a mistake, Galileo just wouldn't make the rearrangement and he would not be forced to look foolish.

The discovery was a good one, and Galileo rearranged the letters to *Cynthiae figuras aemulatur Mater Amorum*. This is Latin for "The Mother of Love imitates Cynthia's shape." The "Mother of Love" is, of course, Venus, and Cynthia is a poetic name for the Moon. What Galileo was saying then was, "Venus changes its shape just as the Moon does."

The Moon, which is Earth's satellite, changes position with respect to the Sun as it circles the Earth. From Earth, we therefore see different portions of its surface lit by the Sun. Since we only see the lit portion, the Moon takes on different shapes or "phases" as it circles the Earth (see Figure 3).

When the Moon is between Earth and Sun, in position A, the Sun shines on the far side of the Moon and the side toward us is dark. We don't see the Moon at all, and it is the "new Moon." On the other hand, when the Moon is on the opposite side of us from the Sun, in position E, the Sun shines over Earth's shoulder (so to speak) and the side we see is completely sunlit. The Moon is then a perfect circle of light, a "full Moon."

As the Moon moves around the Earth from position A to B to C to D to E, the lighted portion of its face slowly expands, as seen from Earth, going from new Moon to a crescent Moon, to a half Moon to a gibbous Moon (GIB-us), which is partway between half and full, to a full Moon. Then, as the

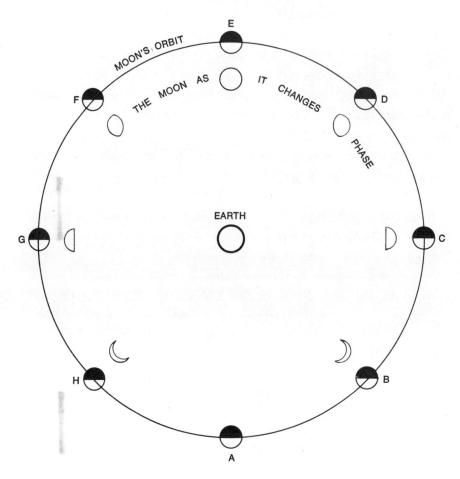

Figure 3
THE PHASES OF THE MOON

Moon goes past the opposite point in the sky and moves closer to the Sun, from position E to F to G to H and back to A, the changes repeat in the opposite direction—from full Moon, to gibbous Moon, to half Moon, to crescent Moon, and back to new Moon. In every revolution of the Moon around the Earth, we see the full cycle of phases.

This doesn't happen in the case of a planet that circles the Sun in an orbit that is larger than our own. In Figure 4, Mars

Figure 4
THE PHASES OF MARS

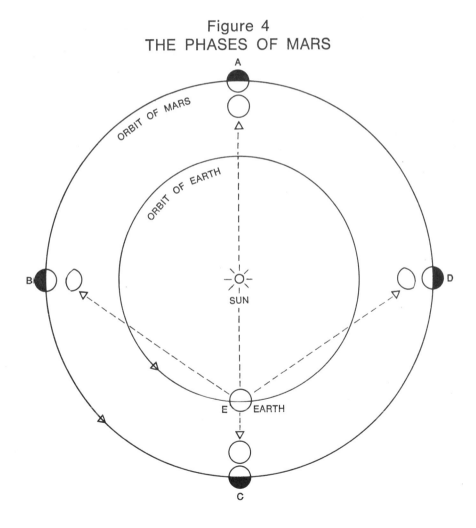

is shown in four positions in its orbit. In each of these positions, the side of Mars toward the Sun is lit up, naturally, while the side away from the Sun is in darkness.

From Earth at position E in its orbit, we see the lighted side of Mars in each case, so that Mars is always "full" or nearly so. When Mars is in positions B and D, we can see a little beyond the lighted side and Mars is then slightly gibbous, but that is as far as it ever gets from full.

For Venus, which orbits the Sun closer than Earth does, the situation is not like that of Mars, but is more like that of the Moon, as we can see in Figure 5.

When Venus is at superior conjunction (position F), we on Earth (at position K) can look past the Sun and see its lighted side. Venus is then in the "full" phase. (Actually the Sun drowns out Venus at this time, but if we could see it, that's what we would see.)

At inferior conjunction (position A), the Sun shines on the side of Venus away from us and it is Venus's dark side that is turned toward us. Venus is then in its "new" phase, and even if we weren't staring directly at the Sun when we tried to see Venus at inferior conjunction, we still wouldn't see it.

If we imagined Venus at inferior conjunction to begin with and watched it move in its orbit, we would see a little of the lighted side, then a little more, then still more and so on. Beginning with new Venus at position A, we would see a thin crescent Venus at position B, then a thicker crescent at position C; then at position D, when Venus was at its greatest elongation, we would see a half-Venus.

As Venus began to move back toward the Sun we would see a gibbous Venus at position E, and finally, at superior conjunction at position F, we would see a full Venus as I said before.

When Venus passes the superior conjunction and moves

Figure 5
THE PHASES OF VENUS

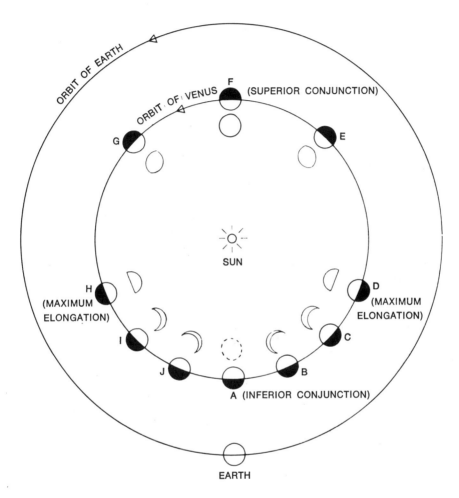

back toward inferior conjunction, through positions G, H, I and J, it passes through all its phases in reverse order.

In this way, Venus goes through all the phases of the Moon in just the order that the Moon does. Of course, Venus is so far away and so small in appearance that we can't make out the phases just by looking at it. Venus looks like a point of light and nothing more, regardless of the phase. With a telescope, however, Galileo saw Venus enlarged and could then make out the phases—and was the first human being to do so.

The importance of the discovery was that Venus's phases followed the pattern that the Copernican view indicated they should, and did so exactly. The older view of the heavens that placed the Earth at the center couldn't explain Venus's phases at all.

This was the final blow to the older theory. Once Galileo discovered Venus's phases, the Copernican system, in which all the planets including Earth revolved about the Sun, had to be accepted.

THE CHANGING SIZE OF VENUS

The phases of Venus are not exactly like the phases of the Moon in every way, however. The Moon circles the Earth and stays pretty nearly at the same distance from us all the time. That means that the full Moon, the new Moon, and all the phases in between are circles or parts of circles of about the same size.

This is not so in the case of Venus. Venus circles the Sun, and when it is at superior conjunction, on the other side of the Sun, it is much farther away from Earth than when it is at inferior conjunction and on this side of the Sun (see Figure 6).

When Venus is at inferior conjunction, it is in the new-Venus phase. Even though it is then as close to us as possible and appears as large as possible, it can't be seen because it is very close to the Sun in the sky. Even if we could manage somehow to blank out the light of the Sun, we still couldn't see Venus, because the side facing us would not be receiving any sunlight. We would be looking at only its night side.

Figure 6
THE CHANGING SIZE OF VENUS

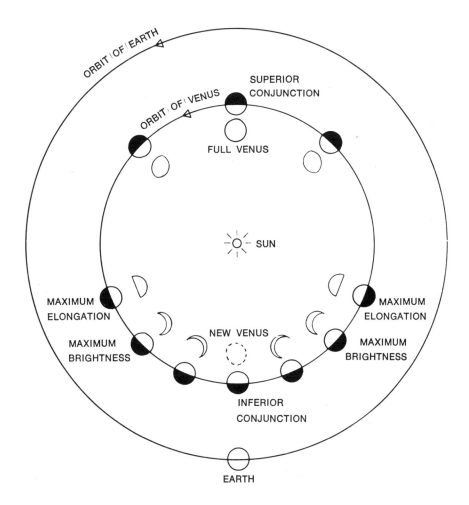

As Venus moves past inferior conjunction in its orbit about the Sun, it begins to show a crescent. The crescent thickens, but as it does so, Venus, in moving in its orbit, also moves farther and farther away, so that the crescent seems to grow smaller as it grows thicker.

The thickening of the crescent sends us more light; the decreasing size sends us less light. For a while, the increase in light resulting from the thickening of the crescent is greater than the decrease in light due to the shrinking of the globe of Venus with distance.

Venus therefore gets brighter and brighter after it passes inferior conjunction and becomes the morning star. It reaches its maximum brightness, however, when it is a fat crescent and *before* it reaches its greatest elongation.

After that point of maximum brightness, the loss in light due to increasing distance overtakes the gain that results from the fact that a greater portion of the globe is sunlit. At the greatest elongation, we have a half-Venus, but the area of the half the globe that is sunlit is a little less than the area of the thick crescent when Venus was closer. For that reason, when Venus is at its maximum elongation it is just a trifle dimmer than it was when it was a little lower in the sky.

After the greatest elongation, Venus continues to increase its distance more rapidly and its brightness continues to diminish, reaching a low point at superior conjunction when it is full Venus. After that, it brightens again as it approaches us and becomes quite bright at maximum elongation as the evening star. Then, as it moves closer to the Sun in our sky, it continues to brighten even further as it comes closer to us, even though the lighted portion of its globe is shrinking. It is at maximum brightness as a thick crescent and then begins to fade off as the crescent thins.

It is a shame, in a way, that the closer Venus comes to us, the less of its globe is in sunlight. Just imagine if, when it was at inferior conjunction, its globe, instead of being entirely dark because we were facing the night side, were entirely lit because, somehow, we saw the day side.

In that case, Venus at inferior conjunction would have a

magnitude of –7.3 and would deliver 17.4 times as much light as Venus ever does now. Such a super Venus would be 1/140 as bright as the full Moon. What a sight that would be!

2

The Orbit
of Venus

ANGULAR MEASURE

When we say that the globe of Venus is much larger at inferior conjunction than at superior conjunction, is there any way of reducing these sizes to numbers in some form of measurement?

We can't actually hold a tape measure up to the sky, but we can make use of "angular measure." The distance around any circle—whether the circle is that of this letter "O" or that of the orbit of a planet revolving about the Sun—can be divided into 360 equal segments, or "degrees." The degree is divided into 60 equal "minutes of arc" and each minute of arc is divided into 60 "seconds of arc."

Since there are 360 degrees to a complete circle, there are 360×60, or 21,600 minutes of arc to a complete circle, and $21,600 \times 60$, or 1,296,000 seconds of arc to a complete circle.

If one were to imagine a circle all around the sky and divide that circle into 720 equal segments, each segment would be half a degree wide, and the Moon and the Sun could each fit snugly into one of those segments. The Moon and the Sun, as they appear in Earth's sky, then, are each roughly half a degree wide, or 30 minutes of arc wide, or 1800 seconds of arc wide.

Even when Venus is at inferior conjunction and is at its largest in appearance, it is still too small for us to estimate its size just by looking at it. If, however, we look at it through a telescope, its size will be magnified and its width can then be easily measured. Knowing the magnifying power of the telescope, we can then calculate the width of Venus as it appears to the unaided eye.

It turns out that Venus's globe appears to be of a size such that if we imagined a number of globes of exactly that size stretched across the Moon as *it* appears to us, it would take about 30 Venuses to stretch from side to side. That means that the apparent width of Venus at its largest is about 1 minute of arc (see Figure 7).

The Moon isn't always at precisely the same distance from us. The distance varies slightly, and when the Moon is closer to us it is a little larger in appearance; when it is farther from us it is a little smaller in appearance. The same is true, to a much greater extent, in the case of Venus.

The angular diameters of the Moon and of Venus at maximum and minimum size are given in Table 2.

Table 2
APPARENT DIAMETERS

Object	Apparent Diameter	
	Seconds of Arc	Venus at Minimum = 1
Venus (minimum)	9.9	1.0
Venus (maximum)	64.5	6.5
Moon (minimum)	1,761	178
Moon (maximum)	2,010	203

Figure 7
APPARENT SIZE OF VENUS

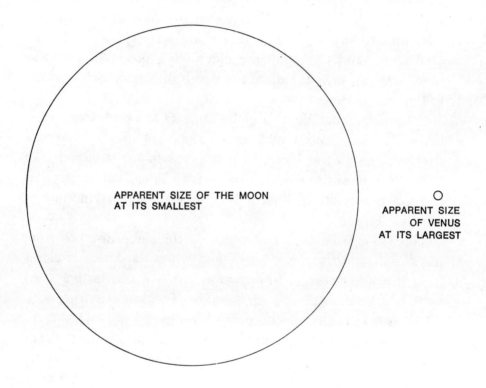

APPARENT SIZE OF THE MOON
AT ITS SMALLEST

○
APPARENT SIZE
OF VENUS
AT ITS LARGEST

Even at its largest, Venus is only about 1/28 as wide as the Moon at its smallest, which is why Venus always appears as a dot of light and why the phases can never be made out.

Brightness, however, doesn't depend so much on the apparent width of a body as on its apparent area. The Moon is not only at least 28 times as wide as Venus from side to side; it is also at least 28 times as wide from top to bottom. For this reason, there is a much larger difference in apparent area between the Moon and Venus than in apparent diameter (see Table 3).

Table 3
APPARENT AREA

Object	Apparent Area	
	Square Seconds of Arc	Venus at Minimum = 1
Venus (minimum)	77.0	1.0
Venus (maximum)	3,268	42.4
Moon (minimum)	2,436,000	31,600
Moon (maximum)	3,173,000	41,200

But then, neither is it the area of a globe alone that counts. How much of that area is sunlit? If we deal with the full Moon we know that all of the area is sunlit, and since the area doesn't change much with phase, we know that the full Moon has to be the time when the Moon is brightest.

When Venus is at the full, its globe is small, however. At maximum brightness only a crescent may be sunlit, but the area of that crescent is larger than the entire globe of full Venus (see Table 4).

Notice that the sunlit area of the Moon is up to 17,000 times as great as the sunlit area of Venus (as seen from Earth). Yet the full Moon isn't nearly that much brighter than Venus at its brightest. This means that each square second of arc of Venus's surface is considerably brighter than each square second of arc of the Moon. I'll explain why later in the book.

The use of angular measure can help us in other ways (see Figure 8).

If 360 degrees measures the complete circuit of a circle, then half a circuit is 180 degrees. Thus, the distance from a given point on a horizon, all the way around the horizon to a point

Table 4
APPARENT AREA OF SUNLIT SURFACE

Object	Apparent Area of Sunlit Surface	
	Square Seconds of Arc	Venus at Maximum Brightness = 1
Venus (full)	77	0.4
Venus (maximum brightness)	188	1.0
Moon (full, at farthest)	2,436,000	12,957
Moon (full, at nearest)	3,173,000	16,877

on the exactly opposite side is 180 degrees. The distance from the point on the horizon, up to the zenith and down again to the opposite point is also 180 degrees.

The distance from any point on the horizon to the zenith is half the distance from horizon to horizon and is 90 degrees. From any point on the horizon halfway to the zenith is 45 degrees.

The maximum elongation of Venus is 47 degrees, or just a trifle over halfway to the zenith. The evening star is 47 degrees up from the western horizon at sunset when it is at maximum elongation, and the morning star is 47 degrees up from the eastern horizon at sunrise when it is at maximum elongation.

PERIOD OF REVOLUTION

It takes time for a planet to make one complete revolution about the Sun. If we imagine the Earth in such a position that a line connecting it with the Sun could be extended onward

Figure 8
ANGULAR MEASURE

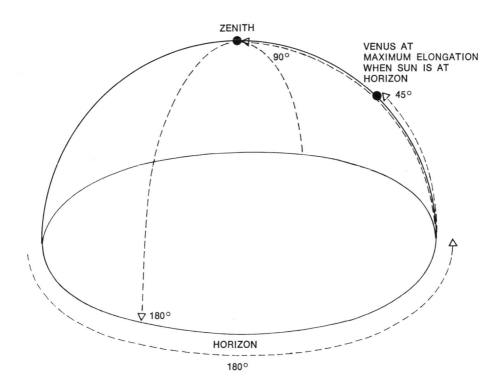

ZENITH

90°

VENUS AT
MAXIMUM ELONGATION
WHEN SUN IS AT
HORIZON

45°

180°

HORIZON

180°

in the direction of a certain star, then it would take just one year for the Earth to make one complete circuit in its orbit and come back to that line.

As the Earth goes around the Sun, its changing position causes it to see the Sun against a changing background in the sky (see Figure 9). Of course, we can't actually see the stars near the Sun, since when the sun is shining all the stars are blanked out. However, astronomers know the starry map of the sky exactly, and from the appearance of the sky at any particular time of the night they can calculate the exact spot in the sky in which the Sun is located.

Figure 9
THE SUN'S APPARENT MOTION

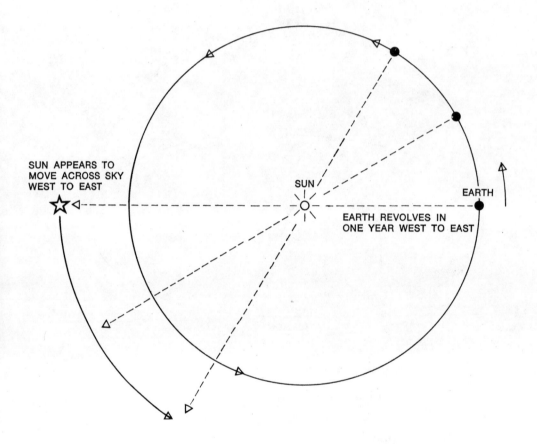

The Sun seems to move against the stars, from west to east, at a speed of just about 1 degree each day. It is possible that that is one reason why the circle was divided into 360 degrees, since with a motion of 1 degree each day, the Sun makes a circuit of the sky in about 360 days.

Actually, the progress of the Sun is not quite 1 degree per day but is about 0.98565 degree per day, and it takes it about

365.24 days to make a complete circuit of the sky. It might have been better to divide all circles into 365 equal parts for that reason, but the ancients who started the practice found 360 a much easier number to handle, since it is so easily divisible by so many smaller numbers.

Since Venus is closer to the Sun than Earth is, and travels about it in a smaller orbit, it seems reasonable to suppose that it would take less time for Venus to go about the Sun than it takes Earth. That is so; Venus's "period of revolution" is shorter than that of Earth (see Table 5).

Table 5
PERIOD OF REVOLUTION

Planet	Period of Revolution		
	Years	Months	Days
Earth	1.000	12.00	365.24
Venus	0.615	7.38	224.70

You might think that if Venus is at inferior conjunction at a particular time and we wait, then after 224.70 days it will be at inferior conjunction again.

That, however, is not the way it works. That would be the case only if the Earth were standing still, but it is not. Like Venus, the Earth is moving about the Sun, though Earth has the longer orbit to follow and takes longer to complete its revolution.

If we start with Venus at inferior conjunction, it will come back to the same position in its orbit after 224.70 days, but Earth will have moved onward and will have completed 224.70/365.24 or just a bit over 3/5 of its orbit. Venus has to follow after, gaining on Earth steadily. It finally catches up to Earth only after it has made about 2.6 circuits about the Sun and has gained a lap on Earth (see Figure 10).

To be exact, after Venus is at inferior conjunction, it is 583.92 days or 1.6 years before it is at inferior conjunction again. Still

Figure 10
SYNODIC PERIOD OF VENUS

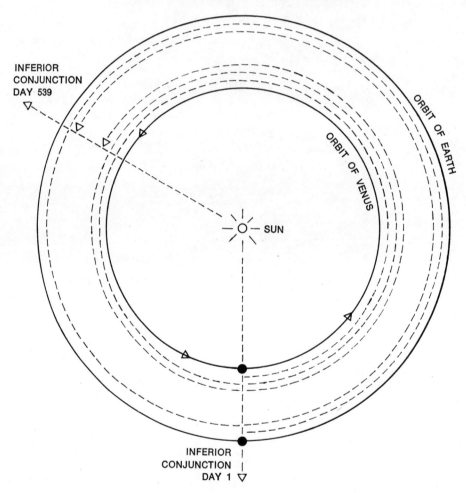

another way of putting it is that there are five inferior conjunctions of Venus, spaced at equal intervals, every eight years.

This 583.92-day period between inferior conjunctions is called the "synodic (sih-NOD-ik) period" of Venus.

NODES AND TRANSITS

As we see Venus move back and forth about the Sun, it moves first in front of the Sun at inferior conjunction; then outward to return behind the Sun at superior conjunction 292 days later; then outward to return in front of the Sun 292 days after that; and so on.

You might suppose that at every inferior conjunction, Venus is *exactly* in front of the Sun, and that at every superior conjunction, Venus is *exactly* behind the Sun as seen from Earth. This is not so.

Imagine an absolutely flat sheet of some exceedingly thin material passing through the center of the Sun and through the Earth. This flat, thin imaginary sheet represents a "plane." It would be Earth's "orbital plane," because the Earth's orbit would be everywhere on this plane, and Earth as it travels about the Sun would always be in the plane.

Venus's orbit also lies in a plane, and that is Venus's orbital plane.

If Venus's orbital plane were the same as Earth's, then Venus's orbit would be lined up exactly with Earth's, and if the orbits were viewed edge-on, Earth's would exactly cover Venus's. In that case, every time Venus was at inferior conjunction, it would be exactly between Earth and Sun, and every time Venus was at superior conjunction, it would be exactly behind the Sun.

However, Venus's orbital plane is not the same as Earth's.

Venus's orbital plane is set at a small angle to Earth's, and that is Venus's "orbital inclination." That orbital inclination is 3.39 degrees (see Figure 11).

Figure 11
ORBITAL INCLINATION

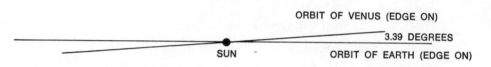

ORBIT OF VENUS (EDGE ON)

3.39 DEGREES

SUN

ORBIT OF EARTH (EDGE ON)

Venus, traveling in its own orbital plane, crosses Earth's orbital plane at a particular point, moves higher and higher, then lower and lower again till it crosses Earth's orbital plane at another point that is at exactly the opposite side of the orbit from the first. Then Venus moves lower and lower, then higher and higher again till it crosses Earth's orbital plane again at the first point. And so on, over and over.

The two points where Venus's orbit crosses Earth's orbital plane are called "nodes."

If Venus reaches inferior conjunction when it is not at a node, it will pass above or below the Sun. If, however, it is at inferior conjunction just when it happens to be at, or very near, a node, it will be in Earth's orbital plane as well as its own. In that case, it will pass exactly between the Sun and the Earth.

When this happens, Venus will be seen to move across the face of the Sun from west to east, appearing as a small black orb that will cover about 1/1000 of the Sun's disc. It will cross the full width of the Sun, if it is exactly at the node, or will cross more or less off-center, if it is a short distance from the node. When Venus crosses the face of the Sun at inferior conjunction, that is called a "transit of Venus."

If Venus passes through the node at one conjunction, it is not going to be at a node the next one or the next. However, since there are five conjunctions in just eight years, then eight years after passing through the node at inferior conjunction, Venus will do so again.

Venus passes through one node on June 7 each year and through the other node on December 9 each year. If inferior conjunction takes place on June 7 in a particular year, it will take place on June 5 eight years later and again there will be a transit.

If there were five inferior conjunctions in *exactly* eight years, then this would happen every eight years without fail. However, the five inferior conjunctions take place in two days less than eight years, so that each eight years, Venus falls farther short of the node at inferior conjunction time. By the time a second eight-year period has passed, Venus is so far from the node that it skims by the Sun, moving a little over it or a little under it and there is no transit.

As the years pass, the inferior conjunction keeps missing the June node by a greater and greater distance and Venus passes over or under the Sun (as viewed from the Earth) with more and more room to spare. At maximum, Venus passes by the Sun at a distance that is several times the apparent diameter of the Sun—so it is quite a miss.

As the years pass, however, the moment of inferior conjunction comes closer to the December node, and eventually it hits that node closely enough to pass directly in front of the Sun, and to do it again eight years later.

The time between the second June transit and the first December transit is 105 years. The time between the second December transit and the first June transit is 122 years. So it is quite possible to live a long life and never experience a Venus transit.

The last two transits, for instance, were on December 8, 1874, and December 6, 1882. The next two will be on June 7, 2004, and June 5, 2012.

There was not and will not be a single transit of Venus in the 1900's.

PARALLAX

Transits of Venus could be important, since they could be used to determine the actual distance of Venus, by the use of "parallax."

You can see how this works if you hold a finger out in front of your eyes at arm's length. If you close one eye, you will see the finger against some object in the background. If you hold your finger steady and close the other eye instead, you will see a change in your finger's apparent position against the background. This change in position is the parallax and, indeed, the word comes from a Greek term meaning "change of position."

If you bring the finger closer to yourself, you will see that the change in position, as you use first one eye and then the other, becomes greater. Parallax changes with distance, becoming greater as the distance becomes smaller, and smaller as the distance becomes greater. By measuring the parallax, you can determine the distance of your finger from your eye.

Using your eyes one at a time, you cannot measure very great distances—only those of several feet at most. For objects farther away than that, the parallax becomes too small to be measured accurately. Fortunately, the shift depends not only on distance, but on the separation of the two points from which the object is viewed. Your eyes are separated by only a few inches, and that isn't much of a "base line." You could use a larger one.

Suppose you drove two stakes into the ground six feet apart. If you viewed an object first from one stake, then from the other, you would increase the amount of the parallax for a given distance, and an object could then be much farther away before the parallax became too small to measure. Your base line might also be greater than six feet—even much greater.

Suppose the Moon is observed at a particular time from a particular position on Earth's surface. At the same time, some-one else observes it from another position hundreds of kilo-meters away. The first observer will see it at a certain distance from a particular star; the second observer will see it at a slightly different distance from that same star. From the shift of position and from a knowledge of how far apart the two observation points are, the distance of the Moon can be calculated. That can be done even without a telescope.

The Moon is by far the nearest of the objects in the sky. The various planets are so far away that their parallaxes are exceedingly small. It was hopeless even to dream of measuring planetary parallaxes until telescopes came into use. Even then, the early telescopes were barely up to the job.

Yet if even one parallax could be measured of one planet, not only would astronomers have the distance of that planet, but they would have the distances of all the other planets as well.

This was because, in 1609, the German astronomer Johann Kepler worked out the first exact model of the planetary orbits.

The ancient Greeks, and even Copernicus and Galileo, thought that the planets moved in perfect circles or in com-binations of perfect circles, whether around the Earth or the Sun. It was Kepler who showed that they move in another kind of curve called an "ellipse," concerning which there will be more to say later.

It was by working with these ellipses that he devised his model and worked out all the distances to scale. He knew, for instance, that Venus was about 7/10 as far from the Sun as Earth was. If someone could work out the parallax of Venus and tell how far it was from Earth at some particular point in its orbit, they could tell from the exact shape of the orbit in Kepler's model how far it was from the Earth at *every* point in its orbit. Then one could calculate the distance of the Sun from Earth and from Venus from the earlier figures, and so on. Unfortunately, the Moon wasn't included in Kepler's model, so knowing its distance didn't help.

Toward the end of his life, Kepler used his model to work out when transits of Venus would take place. Once that was done, it didn't take long for astronomers to see that if the Sun were carefully watched during the transit, the exact time could be noted when Venus first began its trip across the Sun's face, and the exact time when it finished. If the transit were viewed from different spots on the Earth's surface, the exact times of beginning and ending would be slightly different for each, and the exact path taken would be different, too.

This would amount to measuring a parallax, and from these differences the distance of Venus at inferior conjunction could be calculated, and all the other distances in the solar system as well.

This was suggested in 1639 by an English astronomer, Jeremiah Horrocks, who was the first person in history to watch a transit of Venus. Unfortunately, to calculate the distance of Venus you not only need two observers with telescopes, but each one also has to have a good clock. In 1639, clocks good enough to measure time to the second did not exist.

The first clock that could be useful for astronomical measurement was a pendulum clock invented by the Dutch astronomer Christiaan Huygens (HOY-genz) in 1656, and it made all the difference. In 1691, the English astronomer Edmund Halley pointed out that there would be two transits of Venus in the 1700's. He knew he wouldn't survive to see them, but he urged that plans be made to study them thoroughly when the time came. Those transits were going to take place on June 5, 1761, and on June 3, 1769.

Unfortunately, neither transit could be seen well from western Europe. To study them from beginning to end, astronomers would have to travel to Asia or to the Arctic—and in 1761, Great Britain and France were at war and sea travel was unsafe.

However, the Russian scientist Mikhail V. Lomonosov (luh-muh-NOH-suf) observed the Venus transit of 1761, and noticed that as Venus reached the Sun a luminous ring appeared about it. He judged, correctly, that this meant that Venus had an atmosphere, as Earth had, and that sunlight was gleaming through it. He was the first person ever to see Venus's atmosphere. Unfortunately, Lomonosov wrote about his discovery in Russian. In Russia at the time, hardly anyone was interested in science, and outside Russia no one read Russian. It wasn't generally known that Lomonosov had made the discovery till 1910.

In 1769, the English explorer Captain James Cook observed the transit of Venus from the newly discovered island of Tahiti.

In that year, too, an astronomer in Great Britain's American colonies, David Rittenhouse, observed the transit through the first telescope ever constructed on the American continents. Like Lomonosov eight years earlier, Rittenhouse observed that Venus had an atmosphere. Rittenhouse reported it in English, so the world of science learned of it and he got the credit for the discovery.

These transits were used to determine the parallax of Venus, though not very accurately. The figures did fit, however, with those worked out in 1670 by the Italian-French astronomer

Giovanni D. Cassini (ka-SEE-nee) who had, at that time, worked out the parallax of Mars.

Between the figures on Mars and Venus, astronomers had a pretty good notion of the distances of the Sun and the planets from Earth and from each other. In 1835, the German astronomer Johann Franz Encke (ENK-uh) took all the Venus transit figures and worked out the most probable distances from them. For half a century those were the best figures there were.

Finally, the time came for the two transits of the 1800's, one on December 8, 1874, and the next on December 6, 1882. Astronomy had advanced in the preceding century; instruments now were of much better quality and it was far easier to travel. Too, it was peacetime.

Particularly interested was the British Astronomer Royal, George Biddell Airy, who wanted to get figures that were better than Encke's. He put together the best instruments and a team of the best astronomers, trained them thoroughly for the kind of investigations they would have to make—and then it all failed.

Venus's atmosphere made the moment of contact a rather fuzzy and uncertain one. Astronomers were able to get figures, but not very good ones. Despite Airy's enormous efforts, he could not improve on Encke's figures.

DISTANCE AND ECCENTRICITY

Well, then, what are the distances of Venus and Earth from the Sun? The values are given in Table 6—not those of Encke or Airy, but the values as corrected somewhat by new methods of determination worked out in the past century.

The distances are given both in kilometers and miles in this table and throughout the book. Kilometers are used all over

Table 6
DISTANCE OF VENUS FROM THE SUN

Planet	Average Distance from the Sun		
	Kilometers	Miles	Earth to Sun = 1
Venus	108,200,000	67,200,000	0.723
Earth	149,600,000	92,960,000	1.000

the world, but miles are still used in the United States. One kilometer equals about 5/8 of a mile.

Once these distances are known, we can tell the average distance of Venus from Earth at inferior conjunction and at superior conjunction (see Table 7).

Table 7
DISTANCE OF VENUS FROM EARTH

	Average Distance from Earth	
	Kilometers	Miles
At Inferior Conjunction	41,400,000	25,800,000
At Superior Conjunction	257,800,000	160,200,000

Venus seems to be a long way from Earth, but to astronomers it is almost our next-door neighbor. The Moon, which is Earth's satellite, is closer and is only 384,321 kilometers (238,857 miles) from Earth on the average. Except for the Moon, however, no large body ever comes closer to Earth than Venus does. Venus's distance of 41,400,000 kilometers at inferior conjunction puts it only 107.7 times as far away as the Moon.

Notice that in Tables 6 and 7, it is the "average distance" that is given. If the Earth and Venus moved around the Sun in orbits that were perfect circles, then each planet would always be exactly the same distance from the Sun at all times. In that case, Venus would always be the same distance from the Earth at every inferior conjunction and superior conjunction (except for very slight differences that result from the fact that the two orbits are not quite in the same plane).

However, the paths taken by Venus and Earth about the Sun (and, for that matter, by the Moon about the Earth) are not circles, but ellipses, as Kepler showed in 1609.

An ellipse looks like a flattened circle that is exactly alike on both ends. In a circle, any line from edge to edge passing through the center is a "diameter," and all the diameters of a particular circle are equal in length. In an ellipse, the diameters are of different lengths (see Figure 12).

The longest diameter of an ellipse runs from one narrow end to the other and is called the "major axis." The shortest

Figure 12
CIRCLE AND ELLIPSE

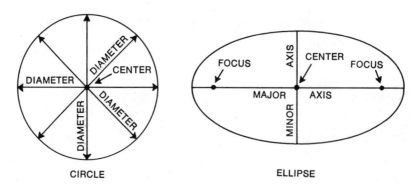

diameter is the "minor axis." The two axes cross at right angles—
that is, if one goes horizontally, the other goes vertically. The
two axes cross at the center of the ellipse.

On the major axis of the ellipse are located two points called
"foci" (FOH-sigh). Each one of them is a "focus" (FOH-kus).
The foci are on opposite sides of the center and at equal distances from it. The foci are located in such a way that if a
straight line is drawn from one focus to any point on the ellipse,
and from that point to the other focus, the sum of the lengths
of the two straight lines is always the same. The sum is always
equal to the length of the major axis, too.

Ellipses can be of different shapes, depending on how flattened they are. The more flattened an ellipse is, the narrower
the ends, and the longer the major axis is compared to the
minor axis. And the more flattened an ellipse is, the farther
the foci are from the center and the closer to the ends (see
Figure 13).

If the foci are only 1/100 of the way from the center to the
ends, we say the "eccentricity" (from Latin words meaning
"from the center") is 0.01. With such a small eccentricity, you

Figure 13
DIFFERENT ELLIPSES

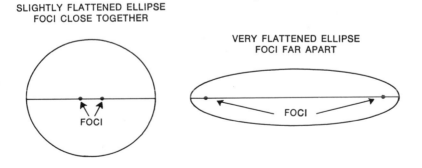

SLIGHTLY FLATTENED ELLIPSE
FOCI CLOSE TOGETHER

VERY FLATTENED ELLIPSE
FOCI FAR APART

FOCI

FOCI

can't notice the flattening and the ellipse looks just like a circle to the eye. (For a circle, the foci are exactly at the center and the eccentricity is 0.)

If the foci are halfway from the center to the ends, the eccentricity is 0.5 and the ellipse looks like an egg that has the same curve on both sides. If the foci are nine tenths of the way from the center to the end, the eccentricity is 0.9 and the ellipse looks rather like a cigar.

Kepler showed that the Sun is located at one of the foci of the ellipse described by the planetary orbit. (And the Earth is located at one of the foci of the Moon's orbit.)

The eccentricities of the orbits of Earth, Venus, and the Moon are given in Table 8. As you see, all three orbits are ellipses of

Table 8
ORBITAL ECCENTRICITY

Orbit	Eccentricity
Venus about Sun	0.0068
Earth about Sun	0.0167
Moon about Earth	0.055

low eccentricity, particularly that of Venus. Of all the planets, in fact, Venus has the orbit of lowest eccentricity, the one that is most nearly a circle.

Since orbits are not exactly circular and since the Sun (or Earth, in the case of the Moon) is not at the center of the orbit, the distances of the planets from the Sun (or of the Moon from the Earth) change from point to point in the orbit.

The Moon is nearest the Earth, for instance, when it is at the end of the major axis of its orbital ellipse that is on the side of

Figure 14
PERIGEE AND APOGEE

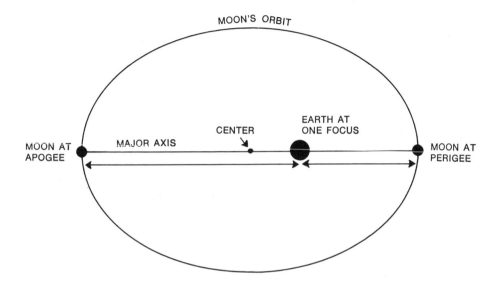

MOON'S ORBIT

CENTER

EARTH AT
ONE FOCUS

MOON AT
APOGEE

MAJOR AXIS

MOON AT
PERIGEE

Note: This diagram is exaggerated to make the difference in distance
clear. In the real orbit, the distance between focus and center
is much smaller than shown in the diagram.

the focus at which the Earth is located (see Figure 14). That point is called the "perigee" (PEHR-ih-jee) from Greek words meaning "near the Earth."

Once the Moon passes perigee on its journey around the Earth, it moves farther and farther from the Earth until it reaches the other end of the major axis. That is the farthest point from the Earth and it is called the "apogee" (AP-uh-jee) from Greek words meaning "away from the Earth."

The different distances of the Moon from the Earth are given in Table 9. It is not a very great change, for the Moon's orbit is not very eccentric, but it is a noticeable one.

The situation of the planets with respect to the Sun is the

Table 9
DISTANCE OF MOON FROM EARTH

	Distance		
	Kilometers	Miles	Average Distance = 1
At Perigee	356,334	221,426	0.927
At Average Distance	384,321	238,857	1.000
At Apogee	406,610	252,667	1.058

same as that of the Moon with respect to the Earth. The closest approach of a planet to the Sun is the "perihelion" (PEHR-ih-HEE-lee-on) which means "near the Sun"; the farthest distance is the "aphelion" (uh-FEE-lee-on) or "away from the Sun." Table 10 gives the different distances of the Earth and Venus from

Table 10
CHANGING DISTANCE FROM THE SUN

	Distance		
	Kilometers	Miles	Average Distance = 1
Venus at Perihelion	107,500,000	66,800,000	0.994
Venus at Average Distance	108,200,000	67,200,000	1.000
Venus at Aphelion	109,000,000	67,700,000	1.007
Earth at Perihelion	147,100,000	91,400,000	0.983
Earth at Average Distance	149,600,000	93,000,000	1.000
Earth at Aphelion	152,100,000	94,500,000	1.017

the Sun. The changes are even less, proportionately, than in the case of the Moon.

Suppose that the inferior conjunction of Venus comes when Earth is at perihelion and is closest to the Sun. At that time Venus is not very far from its perihelion, however, and is therefore a bit farther away than it might be from Earth. Nevertheless the distance between Venus and Earth is then distinctly less than average and is 39,300,000 kilometers (24,400,000 miles). If the Moon should happen, at that time, to be at apogee, then Venus would be, at that moment, only 97 times as far away from Earth as the Moon was.

Once we know the distances of Venus and Earth from the Sun, we can easily calculate the length of each orbit. Since we also know the length of time it takes each to complete one turn of the orbit, we can calculate how rapidly each travels in its journey around the Sun. These figures are given in Table 11.

Table 11
ORBITAL LENGTHS AND SPEEDS

Planet	Orbital Length		Average Orbital Speed	
	Kilometers	Miles	Kilometers per Second	Miles per Second
Venus	679,000,000	421,000,000	35.0	21.8
Earth	939,000,000	584,000,000	29.8	18.5

You can see that if Earth traveled at the same speed about the Sun that Venus does, the length of Earth's year would be only 310.5 days. Earth's year is so much longer than Venus's year not only because Earth travels a longer path, but because it travels over that longer path more slowly.

There is no mystery about why Earth travels more slowly than Venus. Both are in the grip of the Sun's gravity, which whips them on in their orbits. Since Venus is closer to the Sun, it feels the Sun's gravitational influence more intensely, and moves more quickly. In fact a planet travels somewhat more quickly at perihelion, when it is closer to the Sun, than at aphelion, when it is farther away (see Table 12).

Table 12
CHANGING ORBITAL SPEEDS

	Orbital Speed	
	Kilometers per Second	Miles per Second
Venus at Perihelion	35.2	21.9
Venus at Aphelion	34.8	21.5
Earth at Perihelion	30.3	18.8
Earth at Aphelion	29.3	18.2

3
The Size of Venus

DIAMETER AND MASS

Once the distance of Venus is known and its apparent size, as seen at that distance, is also known, then there are well-known mathematical techniques for determining how large an object must be to appear at that size at that distance.

Of course, Venus's atmosphere confuses the issue a bit, and until recently we lacked the techniques for locating the position of its actual surface and calculating the diameter of its solid core. This is something we can do without trouble for the Earth, where we live at the bottom of the atmosphere, and for the Moon, which has none. Nevertheless, this has now been done for Venus, and its diameter as compared with that of the Earth and the Moon is given in Table 13.

Once the diameter of Venus is known, then its surface area and volume can also be determined and compared with the corresponding figures for the Earth and Moon (see Tables 14 and 15).

As you see, Venus is almost the twin sister of Earth in size. No other object in the solar system is as close to the Earth in size as Venus is. And both planets are considerably larger than our Moon (see Figure 15).

Table 13
DIAMETER

	Diameter			
	Kilometers	Miles	Earth = 1	Moon = 1
Earth	12,756	7,927	1.000	3.669
Venus	12,112	7,526	0.950	3.484
Moon	3,476	2,160	0.273	1.000

Table 14
SURFACE AREA

	Surface Area			
	Square Kilometers	Square Miles	Earth = 1	Moon = 1
Earth	511,000,000	197,000,000	1.000	13.40
Venus	460,000,000	178,000,000	0.902	12.11
Moon	40,000,000	15,400,000	0.078	1.00

Table 15
VOLUME

	Volume			
	Cubic Kilometers	Cubic Miles	Earth = 1	Moon = 1
Earth	1,086,000,000,000	261,000,000,000	1.000	49.4
Venus	930,000,000,000	223,000,000,000	0.856	42.3
Moon	22,000,000,000	5,280,000,000	0.020	1.0

Figure 15
SIZE OF VENUS

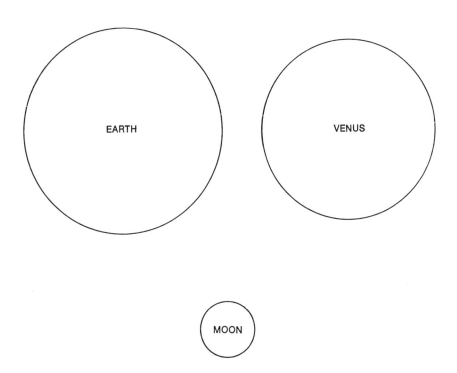

Just the same, the twinship is not total. Earth is distinctly the larger of the two planets. Earth's surface area is larger than that of Venus by 51,000,000 square kilometers (19,000,000 square miles), and that is equal to the area of the continents of Asia and Australia combined. In fact the surface area of the Earth is 10,000,000 square kilometers (3,600,000 square miles) larger than the surface area of Venus and the Moon combined.

And if we imagined a hollow shell the size of Earth and dropped Venus into it, there would be enough room left over within the shell for seven objects the size of the Moon.

Two different objects can have the same size and yet have

very different masses. Thus, an iron sphere the size of a basketball has a much greater mass than a basketball. You can toss
a basketball with ease, but you couldn't even lift that same
basketball if it were made of solid iron.

On the surface of the Earth it is easy to determine the mass
of some object by weighing it by any number of methods. How,
though, can you tell the mass of a large object like the planet
Venus that is millions of kilometers distant?

One way is to measure the gravitational field instead. Every
object that has mass possesses a gravitational field, and the intensity of that gravitational field depends on the mass. If you
can measure the intensity of the gravitational field, you can
calculate the mass.

One way of measuring the intensity of the gravitational field
of a planet is to study the satellites of that planet. A satellite at a given distance from a planet will travel about the
planet at a speed that is related to the gravitational intensity.
The more rapidly the satellite completes a revolution about the
planet, the greater the mass of the planet.

Most planets have satellites, so, in general, it is not hard to
determine planetary masses. Venus, however, is an exception
in this respect. Earth has a companion of considerable size
shining softly in its skies, but Venus is alone and has no satellite
from whose motion we can calculate the planet's mass.

Venus does have some slight gravitational effect on the motions of Earth, and its mass can be calculated, though rather
fuzzily, from that. In recent years, however, human-made objects have passed near Venus, and from the change in the path
of those objects as they passed the planet at a particular distance, the intensity of Venus's gravitational field and, therefore,
its mass can be calculated quite precisely (see Table 16). Mass
is measured by scientists in kilograms, one kilogram being equal
to about 2.2 pounds.

Table 16
MASS

	Mass		
	Trillion Trillion Kilograms	Trillion Trillion Pounds	Earth = 1
Earth	5.976	13.175	1.000
Venus	4.870	10.736	0.815

Once we have the mass, then, since we know the volume of Venus we can work out its density; that is the amount of mass it has for every cubic meter, on the average. (A cubic meter is equal to about 1.3 cubic yards.) The density is given in Table 17.

As you see, the density of Venus is almost equal to that of the Earth, and that tells us something at once about the general chemical makeup of our sister planet.

There are three kinds of common substances in the universe. There are metals—chiefly iron, which has a density of about 7,900 kilograms per cubic meter. There are rocks, with an

Table 17
DENSITY

	Density		
	Kilograms per Cubic Meter	Pounds per Cubic Yard	Earth = 1
Earth	5,518	3,275	1.000
Venus	5,200	3,086	0.942

average density of about 3,500 kilograms per cubic meter. And there are light substances or "ices" such as water, ammonia, and hydrogen, with densities of 1,000 kilograms per cubic meter or less.

Earth has very little of the light substances, but is made up chiefly of a molten core (probably mostly iron) and a rocky mantle around it. That is why the average density of the Earth falls between that of metal and rock. Since Venus has an average density almost that of Earth, it, too, must be made up of a molten core of iron and a rock mantle. In Venus, the molten core makes up just a trifle less of the planet and the mantle a trifle more, and that would account for the slightly smaller overall density of Venus.

Since Venus's mass is 0.815 times that of Earth, Venus produces a gravitational field only 0.815 times as intense as Earth's. At equal distances from the two planets, a particular object would feel only 0.815 times the pull toward Venus that it would feel toward Earth.

If the object were on Venus's surface, however, it would be a little closer toward Venus's center than it would be toward Earth's center if it were on Earth's surface. After all, Venus is a slightly smaller world. This increases the pull of Venus's gravitational field so that the surface gravity is a little higher than you would expect it to be from Venus's mass (see Table 18).

If you were transported to Venus somehow and placed on its surface, you would feel a trifle lighter, perhaps, just at first, but you would quickly get used to it and then feel quite normal.

It is possible to break away from Earth's gravity, if one goes speedily enough. If a rocket ship attains the necessary speed and is beyond the atmosphere, it can shut off its rocket engines, for it will drift outward forever and never return to Earth. (Of course, it will move into orbit about the Sun and might some-day collide with Earth, but that would be an accident and not the same thing.) The lowest speed at which this takes place is the "escape velocity." Venus, with a slightly less intense gravitational field than Earth, has a slightly lower escape velocity (see Table 19).

Table 18
SURFACE GRAVITY

	Surface Gravity (Earth = 1)	Weight of 60-kilogram Person	
		Kilograms	Pounds
Earth	1.00	60.0	132
Venus	0.88	52.8	116

Table 19
ESCAPE VELOCITY

	Escape Velocity	
	Kilometers per Second	Miles per Second
Earth	11.2	7.0
Venus	10.3	6.4

CLOUDS

Since Venus, in size and all that goes with that, seems to be very much Earth's twin, it would be interesting to know whether its surface is like Earth's. Does Venus have continents and oceans, lakes and rivers? Does it, above all, bear life?

Venus certainly has an atmosphere, as Earth does, and that is a hopeful sign, since a world without one (such as the Moon) could scarcely be expected to have surface water, or life as we know it.

Considering that Venus is so close to us—closer than any other planet—we might expect astronomers to be able to learn a good deal about it.

Here, however, it is almost as though Nature is conspiring against astronomers. In the first place, when Venus is at its nearest, at inferior conjunction, only its dark side is facing us and it is, in any case, too near the Sun to be easily visible.

As Venus passes inferior conjunction, more and more of the side facing us becomes sunlit, but at the same time the planet recedes farther and farther from us. By the time it is full Venus, it is at superior conjunction and as far from us as possible (and, in any case, again very close to the Sun). Then, as it passes superior conjunction and approaches us, closer and closer, less and less of the side visible to us is sunlit.

In other words, astronomers have a choice of evils. They can settle for seeing it at close quarters with only a small portion of it sunlit, or for seeing nearly all of it but at very long distance. Either way it is difficult to see much.

Nor is that the worst of it. It early became apparent that wherever and whenever Venus was viewed, nothing was ever seen but blank whiteness over any portion of the globe lit by the Sun.

Clearly, Venus's atmosphere was filled with an unbroken layer of cloud that never lifted, and there was no way in which astronomers' telescopes could penetrate it.

The clouds of Venus explained something. They helped show why Venus appeared as bright as it did in our sky.

All the bodies of the solar system, other than the Sun itself, shine when viewed from a distance only because they reflect light from the Sun.

Not all materials reflect light equally well, however. Rocky materials, for instance, reflect light rather poorly. They absorb most of the light that falls on them. The Moon has no atmosphere, so it shines by the light reflected by its rocky surface. The Moon is bright, to be sure, but it is considerably less bright than it might be. If the Moon reflected all the light that fell on it, it would be about 16 times as bright as it is.

Air reflects more light than rock does, and clouds in particular reflect a great deal of light. Earth's atmosphere is pretty cloudy, so it reflects a greater fraction of light than the Moon does. Venus is completely clouded over, and it reflects a still greater fraction of the light that falls on it.

The fraction of the light falling on a body that is reflected by it is called its "albedo" (al-BEE-doh), from the Latin word for "white." The albedo of Venus is given in Table 20.

Table 20
ALBEDO

	Albedo
Earth	0.36
Moon	0.06
Venus	0.61

Venus's high albedo is an important reason why it is so bright in our sky, and for that we may thank its cloud layer. The price for Venus's brilliance is high, however. As time went on, astronomers learned a great deal about many of the planets, but Venus, the nearest of them all, remained a complete mystery in many respects.

As the 1800's wore on, for instance, astronomers were able to decide how rapidly each planet spun on its axis and whether the axis was tilted to the vertical and by how much—but none of this was true for Venus.

In order to determine how fast a planet rotates, one has to find some noticeable spot on its surface and watch as that spot moves about the planet and returns to the original place. From the direction of motion of that spot, one can then determine the tip of the axis.

There was no use trying this on Venus. There were no spots to be seen, just blank whiteness. Every once in a while some astronomer thought he saw a spot and then announced Venus's period of rotation, but no two astronomers agreed. Even in the 1950's, when the period of rotation was known, or thought to be known, for every other planet, Venus's period of rotation remained a mystery.

Then, too, astronomers managed to prepare interesting maps of Mars's surface by the end of the 1800's. Venus was larger than Mars and was closer, but there were no maps of Venus's surface—not unless you consider a blank patch of whiteness a map.

WATER

Of course, there was nothing to stop astronomers from *guessing* what Venus might be like under the clouds.

For one thing, they naturally assumed the clouds were composed of water vapor. Venus was Earth's twin in so many ways, it seemed unlikely that it wouldn't be watery and cloudy as Earth is. The fact that Venus was more cloudy than Earth seemed natural, too, because Venus, being closer to the Sun than Earth, would be warmer. That meant that on Venus more of the oceans would evaporate in the Sun's warmth, more water vapor would rise into the upper atmosphere, and more clouds would form.

Knowing Venus's distance from the Sun, it is not hard to calculate how large the Sun would appear as seen from Venus and how much light and heat Venus would receive from the Sun as compared to Earth (see Table 21).

Table 21
THE SUN VIEWED FROM VENUS

	Sun		
	Apparent Diameter (minutes)	Apparent Area (square minutes)	Light Received (Earth = 1)
Earth	31.99	803.74	1.00
Venus	44.22	1,536.1	1.91

In that respect, the clouds might even be a good thing, for it could be argued that they would surely cut down on the amount of heat absorbed by Venus and that that would keep Venus's surface cooler.

Venus's clouds reflect 0.61 of the sunlight that falls upon the planet, so that if Venus receives 1.91 times as much sunlight as Earth does, then the amount of light and heat that is actually absorbed by the atmosphere and surface of Venus is only 0.74 times as much as Earth receives altogether.

Of course, Earth's partly cloudy atmosphere reflects 0.36 of the light that strikes it, so Earth actually absorbs only 0.64 times as much as it receives altogether. Therefore, the amount of light and heat from the Sun that Venus actually absorbs is only 0.74/0.64 or 1.16 times that which is absorbed by Earth.

That might be taken to mean that Venus is warmer than Earth, but not by very much. It might well be a humid tropical world, possibly very hard to bear in the equatorial regions, but warm in the temperate zones, pleasantly cool on the plateaus, and delightfully mild even in the polar areas.

A great deal would depend, to be sure, on how much water there might be under the clouds. Considering the unbroken thickness of the clouds, perhaps there is a good deal of water on the surface, more than there is on Earth. In that case, Venus's surface might be an unbroken ocean—or an almost unbroken one, with a few small islands here and there, representing the tops of mountain peaks rising from the ocean floor.

On the other hand, maybe there isn't much water, but what exists is spread out evenly over low-lying land so that Venus is a world of unbroken marshes and bogs.

One particularly popular view of what Venus might be like under its cloud cover depended on a theory advanced in 1799 by the French astronomer Pierre Simon de Laplace (lah-PLAHS).

He suggested that the solar system began as a vast cloud of dust and gas which slowly swirled and which, under the influence of its own gravitational field, slowly came together and shrank into a smaller and smaller volume. As it shrank, it rotated faster and faster, in accordance with a natural law called "the conservation of angular momentum." Finally, the cloud was rotating so quickly that material spun away from its equatorial regions, forming a ring of matter that condensed to form a planet.

The cloud continued to shrink and to speed up its rotation,

giving up ring after ring of matter, and forming planet after planet, until what was left condensed to form the Sun.

If this theory were correct, the outermost planet would be the oldest, and the planets would grow younger as one moved in toward the Sun, with the innermost planet being the youngest. And the Sun itself would be younger than any of the planets.

In particular, Laplace's theory seemed to indicate that Venus was younger than Earth.

This fixed itself in the minds of people, and it became very common to think of Venus as a watery planet on which life had developed as it had on Earth, but had not yet reached as advanced a stage in evolution. Venus, in that view, might perhaps resemble Earth in the early days of the dinosaurs.

By 1900, though, astronomers felt that Laplace's theory did not explain all the facts about the solar system and that something better would have to be found. Different theories were suggested as the 1900's wore on, but in all of them all the planets were formed at the same time and all were equally old.

Astronomers now believe that the Earth, the Moon, Venus, and all the other objects in the solar system, including the Sun, are about 4,600,000,000 years old. In particular, Venus is no younger than Earth.

But even if we have to abandon the thought of a primitive Venus, it might still be very watery. That remained in the minds of people into the 1950's. Over and over again, science-fiction stories described Venus as wet, swampy, boggy, as a planet of perpetual rain, or even as a world with a surface that was one big ocean.

In 1954, I myself wrote a novel for young people entitled *Lucky Starr and the Oceans of Venus* in which I described the planet as being covered by one big ocean that was full of life.

That guess seemed all the more reasonable when one year later, in 1955, two American astronomers, Donald H. Menzel and Fred L. Whipple, summarizing what was then known about Venus, argued that it might have a planetary ocean.

Still, whatever the details of Venus's surface and however pleasant its climate might be, there was no doubt that it would have to be a gloomy world by Earth standards. Nighttime on Venus would always be pitch-dark for there would be no satellite in the sky, and it would not be visible if there were. Nor would any of the stars or planets be visible—no one on Venus's surface would ever see Earth shining gloriously in the sky with a magnitude reaching −6.6 at times, far brighter than even Venus appears in our own sky.

Even by day the world would be a uniform gray, with the bright Sun showing up perhaps as nothing more than a hazy dim-yellow spot in the clouds. Venus would certainly be no world for an astronomer.

CARBON DIOXIDE

Not everyone agreed that Venus might be watery. There were some who wondered whether it might not be fairly dry, even completely dry.

To explain those doubts we must go back to 1859, when the German scientist Gustav Robert Kirchhoff (KIRK-huf) showed that hot substances give off certain wavelengths of light that can be detected and identified by an instrument called a "spectroscope."

Each different kind of atom gives off wavelengths characteristic of itself and of no other kind of atom. This means that every element (a substance made up of but one kind of atom) has a spectroscopic "fingerprint," so to speak, that can be used for identification.

Whereas hot elements give off wavelengths of light, cold ones absorb wavelengths, the same wavelengths they would give off if they were hot. These absorptions can also be used to identify elements.

Thus, sunlight given off by the Sun's hot surface contains all the wavelengths of light. That sunlight passes through the gases around the Sun (its atmosphere) and those gases, while hot, are nevertheless cooler than the shining surface of the Sun. The Sun's atmosphere therefore absorbs certain wavelengths.

If the sunlight is analyzed by the spectroscope, the missing wavelengths show up as dark lines across the band of changing colors of light (the "spectrum"). The location of those lines tells astronomers what elements are present in the Sun and in what quantity and at what temperature and so on.

Atoms can combine into groups called "molecules." For example, two hydrogen atoms form a hydrogen molecule (H_2), while two oxygen atoms form an oxygen molecule (O_2). The atoms of one element can also combine with atoms of other elements. Two hydrogen atoms and an oxygen atom form a water molecule (H_2O), while a carbon atom and two oxygen atoms form a carbon dioxide molecule (CO_2).

Molecules can also give off or absorb wavelengths of light, especially wavelengths that are long and low in energy. In the ordinary light we see, red light has the longest waves and lowest energy. Beyond the red light, however, is "infrared radiation," which our eyes are not sensitive to but which can be detected and studied by appropriate instruments. Molecular "fingerprints" are sometimes found in the visible light spectrum, but more often in the infrared spectrum.

When sunlight passes through a planet's atmosphere, the

molecules in the atmosphere absorb various wavelengths of the spectrum. If you know which wavelengths are already absorbed by the Sun's atmosphere, then you can pick out the additional absorptions and blame them on particular molecules in the planet's atmosphere.

This isn't as easy as it sounds, for Earth's own atmosphere absorbs some of the wavelengths, too, and this must be allowed for. When this is done, what is left over can consist of very faint absorptions indeed, and it isn't always easy to be sure what you are dealing with.

It wasn't till 1932 that a molecule in Venus's atmosphere was identified with considerable confidence. In that year, the American astronomers Walter Sydney Adams and Theodore Dunham, Jr., detected carbon dioxide.

That seemed odd, since with all those clouds, it might have seemed that the water molecule should have been detected. Or if Venus is truly Earth's twin, perhaps nitrogen or oxygen (which make up 99 percent of Earth's atmosphere) should have been detected.

Since it was carbon dioxide that was detected, it seemed reasonable to suppose that there was a lot of it in Venus's atmosphere. There is carbon dioxide in Earth's atmosphere, too, but only about 0.035 percent in quantity. If astronomers were looking at Earth's atmosphere from Venus, they would never detect carbon dioxide present in such small quantities.

If Venus's atmosphere is rich in carbon dioxide, that makes it less likely that the planet is rich in life (at least, in life that is similar chemically to that on Earth).

Green plants on Earth build their tissues by combining carbon dioxide with water, using the energy of sunlight to do so. Oxygen is left over and is discharged into the atmosphere. Once, earlier in Earth's history, our atmosphere was probably rich in carbon dioxide, but the activity of green plants consumed that carbon dioxide and replaced it with oxygen.

Animals on Earth get their energy by breathing oxygen and combining it with chemicals in the food they eat, thus forming carbon dioxide again. Both oxygen and carbon dioxide remain in balance, and both plants and animals flourish.

With so much carbon dioxide on Venus, it would seem that there could scarcely be our kind of plant life on the planet or all that carbon dioxide wouldn't be there. And without plant life and the oxygen it produces, animal life of an earthly kind could not exist.

Of course, there might still be primitive one-celled forms of life of the kind Earth possessed before its atmosphere grew oxygen-rich. Or Venus life might be different chemically from Earth life.

There is, however, another and even worse problem.

Suppose Venus, like our Moon, had no atmosphere. As it rotated, each portion of its surface would absorb sunlight as it passed through the day hemisphere and would grow warmer. Then, as each portion passed through the night hemisphere it would radiate the heat into outer space in the form of infrared radiation.

If we imagine Venus starting at a very low temperature, it would gain heat from the Sun by day but wouldn't lose very much of it at night. Each day it would gain more than it lost at night and its overall average temperature would go up.

As the overall average temperature went up, it would absorb somewhat less sunlight by day and would lose more heat in the form of infrared at night. Finally, at some particular temperature, it would radiate as much heat at night as it gained by day. Then its average temperature would remain steady.

If Venus's atmosphere were made up of oxygen and nitrogen, like our own, that wouldn't change matters much. Sunlight passes through oxygen and nitrogen easily, and so does infrared. (Because there's not much absorption, it is hard to detect oxygen and nitrogen by spectroscope, by the way, and that *might* be one reason why those gases weren't detected in Venus's atmosphere early on.)

Carbon dioxide is another matter. The visible light that makes up sunlight passes through carbon dioxide without trouble, but the longer wavelengths of infrared are absorbed. (One reason why Adams and Dunham were able to detect carbon dioxide in Venus's atmosphere.) Even a little carbon dioxide in the atmosphere absorbs enough infrared to prevent the planet from radiating away nearly as much heat as it would in the absence of carbon dioxide. This means that the temperature of the planet rises to a higher temperature than it otherwise would before enough infrared manages to get through the atmosphere to prevent further heating.

Even the small quantity of carbon dioxide in Earth's atmosphere is enough to keep Earth a little warmer than it would otherwise be. If there were no carbon dioxide at all, Earth might be cold enough to move into an Ice Age.

The ability of carbon dioxide to warm a planet is called the "greenhouse effect" because something of the same sort is supposed to happen in a greenhouse. Sunlight enters the greenhouse and heats the air inside, and the heat then finds it difficult to get out again.

If Venus has quite a bit of carbon dioxide in its atmosphere, that, combined with the fact that it is closer to the Sun than Earth is, might mean a large greenhouse effect. Venus could then be considerably warmer than we would expect. In fact, it might be so hot that its surface would be over the boiling point of water. In that case any ocean it had might have boiled away long ago into water vapor and that vapor might have condensed into tiny droplets in the cool upper atmosphere. In short, Venus's cloud layer may be its ocean, and that might be why the planet has so thick and extensive and permanent a cloud layer.

But that creates problems, too. Water on Venus's surface would be protected from the more energetic portion of sunlight by a deep atmosphere that would absorb much of it. If the water existed in fine droplets high in the atmosphere, it would not be so protected.

The portion of visible light that has the shortest waves and is most energetic is violet light. Beyond it, with even shorter waves and more energy, is "ultraviolet radiation." Our eyes are not adapted for seeing it, but they can be damaged by it, and it is the ultraviolet radiation that gives us sunburn.

High in the atmosphere, the energetic action of the Sun's ultraviolet radiation can pull the water molecule apart, forming free oxygen and hydrogen atoms which might combine, in pairs, to form oxygen and hydrogen molecules.

If this happened, the hydrogen atoms or molecules would be too light for Venus's gravitational field to hold on to them. They would escape into space. Oxygen atoms or molecules would be held by Venus's gravitational field but would soon combine with other atoms. In that way, water that existed only in the form of fine droplets in the upper atmosphere would slowly disappear. After 4,600,000,000 years of exposure to sunlight at close distance, it might well be expected that all the water was gone and that Venus was bone-dry.

But in that case, how would one explain the clouds?

One possibility would be that a dry Venus was experiencing an enormous and perpetual sandstorm and the clouds were clouds of dust and grit.

In 1937 the German-American astronomer Rupert Wildt

(Vilt) suggested that some of the water molecules weren't split apart by ultraviolet radiation, but combined with carbon dioxide molecules instead. Such combination, making use of the energy of ultraviolet, could form molecules of formaldehyde, which is made up of one carbon atom, two hydrogen atoms, and one oxygen atom (CH_2O).

Formaldehyde is a gas, but once it is formed, it could combine into molecular groups that would form a white solid. The clouds of Venus, Wildt suggested, might be made up of this solid formaldehyde combination.

Such a dry Venus with clouds that were either dust or formaldehyde could not possibly bear life in any form similar to that which is familiar to us.

In the 1950's, then, astronomers were torn between the notion of a wet Venus and a dry Venus and there seemed no way of making a final decision. It would help if water vapor could be detected in Venus's atmosphere, but apparently it couldn't be.

The wet-Venus astronomers didn't take that as final, however. For one thing, most of the water vapor on Earth is in the bottom few miles of the atmosphere. There is very little in the upper atmosphere. Since all the light we get from Venus comes from the top of its clouds and passes through Venus's upper atmosphere only, we can expect that it would show only very faint signs of water vapor even if Venus were very wet. Those faint signs would be drowned out by the absorption of light by the water vapor in our own atmosphere.

One way of getting around that would be to make spectroscopic studies of Venus from balloons that would lift the instruments high above the lower atmosphere where almost all of Earth's water vapor exists. This was done in 1959 by a team led by an American physicist, John Donovan Strong, and water vapor was indeed detected.

This meant that there was still a possibility that Venus, Earth's twin in size and composition, might also be Earth's twin in being cool and wet. But that was not yet certain, for it might also be hot and dry.

The final decision came in the early 1960's.

4
The Rotation of Venus

TEMPERATURE

A hot, dry Venus or a mild, wet one! Might not the matter be checked by direct measurement of Venus's temperature?

In 1955 two American astronomers, Edison Pettit and Seth Barnes Nicholson, managed to measure the amount of energy in the light reaching Earth from Venus. From this, they could calculate the temperature of that portion of Venus's atmosphere at the top of the cloud layer—the portion that reflects the light. It turned out that the temperature was −38° Celsius (−36° Fahrenheit).

The temperature of the upper atmosphere is no guide, though. That would be cold no matter what the temperature at Venus's surface might be. Our own atmosphere a couple of dozen miles up, above our own cloud layers, is much colder than Earth's surface is.

What had to be done was somehow to penetrate to Venus's surface—and the tools for the job appeared.

Infrared waves can come in varieties with greater and greater wavelengths. When the waves get to be long enough, they are no longer referred to as infrared. The longer-wave radiation is referred to as "microwaves." When the waves are still longer, they are called "radio waves."

Properties change with the lengths of the waves. The longer the waves, the more easily they penetrate matter. The shorter the waves, the more easily they are reflected from matter.

Long radio waves can penetrate solid matter quite easily and are not reflected from ordinary objects (which is why radio waves reach our radio and television sets through the walls of our houses and through other obstacles between ourselves and transmitting stations). Short waves of ordinary light are easily reflected (we see by reflected light), but are easily stopped by matter that is even as insubstantial as mist or smoke.

Intermediate wavelengths, such as those of microwaves, penetrate fog, mist, smoke, and clouds easily, but are as easily reflected by more solid obstacles.

During World War II, scientists learned how to produce a beam of microwaves and send it out so that it would be reflected by airplanes and other obstacles, and the reflected beam received. From the direction of the reflected beam the direction of the airplanes could be worked out, and from the time it took the reflected beam to return, the distance of the airplanes could be worked out.

In this way approaching enemy planes could be detected, located, and fought off before they reached their destinations. The defenders could not be taken by surprise. Great Britain, after developing this device, used it to defeat Germany in the Battle of Britain in 1940. The device was called "radar."

After the war, astronomers began to send beams of microwaves into space where they would be reflected by astronomical bodies. In 1946, for instance, a Hungarian scientist, Zoltan Lajos Bay, sent microwaves to the Moon and detected the reflected beam 2.5 seconds later.

Microwaves travel at the speed of light, a speed that is known quite accurately. From the time that passed between sending the beam and receiving its reflection, the distance of the Moon at that moment could be determined more accurately than ever before.

The distance of the Moon could not be used to work out other distances in the solar system. However, if a beam of microwaves could be reflected from one of the planets, that would give a very accurate measure of the distance of that planet at that time. From that, as explained earlier in the book, all the distances in the solar system could be worked out.

Since Venus was the nearest of the planets, it was the obvious target. Venus was a hundred times as far away as the Moon, of course, and that made it a much more difficult target. It was not until 1958 that Venus was hit by a beam of microwaves and the reflection detected. By use of this technique, the various distances in the solar system came to be known better than ever before.

It turned out, for instance, that the distance from the Earth to the Sun was about 80,000 kilometers (50,000 miles) less than had been thought. That's not much out of a total distance of 149,600,000 kilometers (93,000,000 miles), only 1/20 of one percent, but astronomers were glad to have the correction.

As astronomers learned how to detect long-wave radiation more and more delicately, they found they had reached the point where they could detect the radiations that bodies in space gave off.

Cool bodies, such as the Earth, do not radiate visible light, but they do radiate the lower-energy infrared radiation. In fact, they radiate a whole range of wavelengths right into the radio-wave region. The pattern of wavelengths that is radiated depends entirely on the temperature of the object that is doing the radiating. From the pattern, the temperature can be determined as surely as if we had put a thermometer to the object.

In 1956 a team of American astronomers, headed by Cornell H. Mayer, detected the long-wave radiation given off from Venus's night side and analyzed its pattern. They found that microwaves with a wavelength of 3 centimeters (1.2 inches) were coming from Venus with surprising intensity. There shouldn't be that much radiation of that type coming from any object unless it is about 330° Celsius (625° Fahrenheit). That is well above the boiling point of water, which is 100° Celsius (212° Fahrenheit).

This was the first indication that Venus might be hotter than anyone had supposed, hotter even than those who believed in a hot, dry Venus had thought.

It was indeed a startling finding, but it was not taken entirely seriously. As it had been determined by a new technique, there might be shortcomings in the procedure. Or there might be more than one explanation for the finding. The microwaves might be coming from some hot region of the atmosphere and not from Venus's surface.

What was needed was a closer look—and that was beginning to seem possible.

The very year after the first microwave indication of a very hot Venus, the Soviet Union placed the first artificial satellite into orbit about the Earth, on October 4, 1957. Very soon, the Soviet Union and the United States were competing to see who could make the more spectacular space shots.

In early 1959, both the Soviet Union and the United States sent artificial satellites out into space at velocities greater than the escape velocity. Such satellites were no longer prisoners of Earth's gravity, but went into orbit about the Sun.

They were launched under conditions that carried them

close to the Moon as they took up their orbit. That made them "lunar probes."

Was a lunar probe all that was possible? The nearest large body beyond the Moon was Venus. Might there not be a Venus probe? Might there not be a probe that could automatically receive, study, analyze, and report on the microwaves emitted by Venus at close range?

The first successful Venus probe was Mariner II, launched by the United States on August 27, 1962. It traveled through the vastness of interplanetary space, taking a curved path that carried it 290,000,000 kilometers (180,200,000 miles) in 109 days. Finally, on December 14, 1962, when it was 58,000,000 kilometers (36,000,000 miles) from Earth in a straight line, it skimmed past Venus at a distance of 35,000 kilometers (22,000 miles) above its cloud layer. At that time, Mariner II approached Venus to within less than a tenth of the distance between the Earth and the Moon. If a human eye had been on the probe it would have seen Venus with 900 times the area and 18,000 times the brightness that we see the full Moon here on Earth.

Mariner II was able to measure the microwave radiation from Venus in great detail from various spots on its globe. It showed that the surface of Venus was hotter than even the first microwave measurements on Earth had indicated, and all later probes have confirmed that.

The surface, we are now certain, is hellishly hot all over Venus, near the poles as well as at the equator, and on the night side as well as on the day side. The surface temperature is something like 475° Celsius (890° Fahrenheit), which is more than hot enough to melt tin and lead and to boil mercury. There's no chance for liquid water anywhere, or for life as we know it either.

SLOW ROTATION

Another discovery that Mariner II made was that Venus has no magnetic field.

Earth does have a magnetic field. Earth acts like a gigantic magnet with a north magnetic pole in the arctic area and a south magnetic pole in the antarctic area. That's why magnetized compass needles point north and south.

The Earth's magnetic field bellies out into space all around it, as "lines of force" travel from the north magnetic pole to the south magnetic pole. These lines of force are imaginary lines along which the intensity of the magnetic field is unchanged.

Objects which do not carry an electric charge can cross the magnetic lines of force without being affected by them. Objects which *do* carry an electric charge, however, are deflected by the magnetic lines of force. If the objects are light and do not carry much energy, they are stopped altogether and are trapped. They spiral back and forth along the lines of force and finally leak into Earth's atmosphere near the magnetic poles, where the lines of force curve downward toward the Earth's surface.

The Sun continually sprays electrically charged particles in all directions—particles much smaller than atoms and hence called "subatomic particles." These subatomic particles make up the "solar wind." Some of the particles end up in the neighborhood of the Earth and some of these are trapped by the lines of force. The whole planet is surrounded by them in a shell that is quite close to the Earth on the day side but bellies far out on the night side.

This shell of electrically charged particles encompassing the Earth was originally called the "Van Allen belt," since a team under the leadership of the American physicist James Alfred Van Allen discovered these particles from satellite observations in 1958. Nowadays it is called the "magnetosphere."

Why does the Earth have a magnetic field?

Current thinking credits two facts. First, the Earth has a liquid metallic core that can conduct an electric current. Second, the Earth rotates fairly rapidly. The rotation, it is thought, sets up swirls in the liquid core, and turning electric currents will set up a magnetic field.

What of the fact, then, that Venus does *not* have a magnetic field and therefore does not have a magnetosphere?

From its size and density, astronomers are certain that Venus must have a liquid metallic core. In order to lack a magnetic field, then, it must lack the other necessary factor. It must rotate slowly, so slowly that the liquid core is not set to swirling.

This notion that Venus rotated very slowly was not exactly new at the time of Mariner II. Observations from Earth suggested reasons to believe in a slow rotation. There was the question of Venus's shape, for instance, and what that indicated.

To explain that, we have to understand that any object that moves in a circle experiences a "centrifugal" effect. This term comes from Latin words meaning "to flee the center." The centrifugal effect, as the name implies, tends to force matter away from the center of rotation. You can feel this if you whirl a heavy weight tied to a string. The weight will pull at your finger as you whirl it, and if you whirl it quickly enough the centrifugal effect will break the string. The centrifugal effect increases as the speed of turning increases.

The Earth, or any other spherical body in rotation, turns about an axis, which is an imaginary line passing through the center of the body and emerging at the north and south poles.

The north and south poles are on the axis and don't move in a circle as the Earth rotates, so there is no centrifugal effect there. The farther you move away from the poles, however, the farther the surface of Earth is from the axis and the faster it must move to make a complete turn in twenty-four hours (see Figure 16).

The equator is the line around the Earth exactly halfway

Figure 16
THE ROTATING EARTH

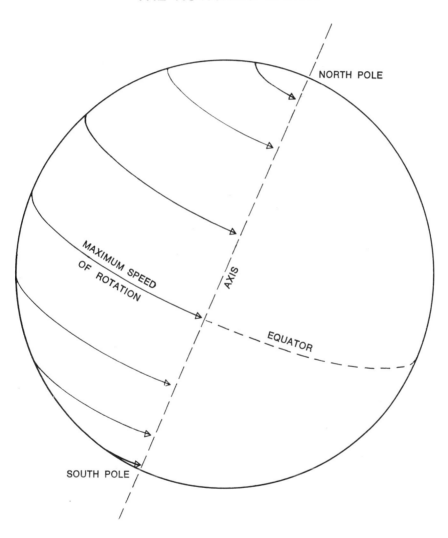

between the poles. It is farthest from the poles and the surface is moving fastest there as the Earth rotates.

Since we know that the Earth has a diameter of 12,756 kilometers (7,927 miles), it is easy to calculate its circumference—that is, the distance all around the Earth at the equator. It comes out to 40,074 kilometers (24,902 miles).

A point on the equator, which must make a complete turn in 24 hours, therefore travels at the rate of 1,670 kilometers per hour (1,038 miles per hour). For a comparison of these figures with those of some other members of the solar system, see Tables 22 and 23.

Table 22
CIRCUMFERENCE

	Circumference		Period of Rotation	
	Kilometers	Miles	Days	Hours
Earth	40,074	24,902	1.00	24
Moon	10,920	6,786	27.32	655.7
Jupiter	449,900	279,600	0.41	9.83
Sun	4,373,000	2,717,400	25.0	601

How do these differences in equatorial speed show themselves? Let us consider Earth first.

The centrifugal effect tends to lift the surface of the Earth away from the center. As the speed of the Earth's surface increases as one moves away from the poles, the lifting effect increases and reaches a maximum at the equator.

The surface of the Earth therefore bulges (very slightly) as

Table 23
EQUATORIAL SPEED

	Equatorial Speed		
	Kilometers per Hour	Miles per Hour	Earth = 1
Earth	1,674	1,040	1.00
Moon	16.65	10.35	0.01
Jupiter	50,850	28,440	30.5
Sun	7,276	4,521	4.35

one leaves the pole, and the height of the bulge is greatest at the equator (see Figure 17). Earth has, in other words an "equatorial bulge." Another way of putting it is that Earth is flattened at the poles.

Consequently, the diameter of the Earth that goes from north pole to south pole (the "polar diameter") is a little shorter than the diameter of the Earth that goes from one point on the equator to the opposite point on the other side of the globe (the "equatorial diameter").

This means that the Earth is not a perfect sphere, in which all diameters would be of the same length. Instead, thanks to its rotation and the centrifugal effect, it is an "oblate spheroid."

The difference in the lengths of the diameters is not great, to be sure. The polar diameter is 12,714 kilometers (7,902 miles) long and the equatorial diameter (which I used in calculating the equatorial speed, of course) is 12,756 kilometers (7,927 miles) long. The difference is 42 kilometers (26 miles), so the bulge is 21 kilometers (13 miles) thick (at its thickest point, on the equator) all around the world.

Figure 17
THE EQUATORIAL BULGE

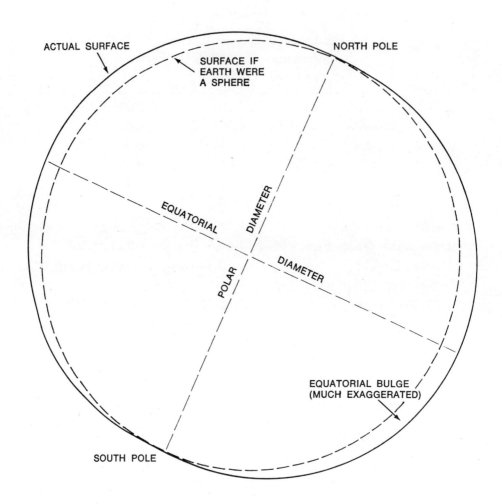

The difference between the two diameters divided by the length of the polar diameter gives a quantity known as "oblateness." For Earth it is 42/12,714 or 0.0033.

This is not much of an oblateness. If you could look at the Earth from a distance it would seem like a perfect sphere. The equatorial bulge would be too small to notice. However, the bulge is there, and it can be measured. For the oblateness of the other globes mentioned in the previous two tables, see Table 24.

Table 24
OBLATENESS

	Oblateness	
		Earth = 1
Earth	0.0033	1.0
Moon	0.0000	0.0
Jupiter	0.0637	19.4
Sun	0.0000	0.0

As you see, Jupiter is much more oblate than Earth is— nearly twenty times as oblate. It is so oblate that if you look at Jupiter through a telescope, you can see that it is not a perfect sphere, and that it bulges at the equator. This is not surprising, since Jupiter's equatorial speed is 30.5 times that of Earth. That the oblateness is only twenty times Earth's can be explained by the fact that Jupiter has a much more intense gravitational field than Earth has, and it is therefore more difficult to lift Jupiter's substance against its own gravitational pull.

The Moon, on the other hand, has such a small equatorial speed that not enough centrifugal effect is produced to lift the Moon's surface a noticeable amount against its own gravitational field, small though it is. There is no measurable equatorial bulge at all, and no oblateness. The Moon is an almost perfect sphere.

The Sun seems puzzling at first glance. Its equatorial speed is over four times that of Earth, and yet it has an oblateness of zero and no detectable equatorial bulge. But then, the intensity of the Sun's gravitational field is so enormous that even an equatorial speed considerably greater than Earth's isn't enough to produce an equatorial bulge.

If, then, a world should prove very nearly a perfect sphere, we must conclude that it is either spinning quite slowly or that it has an enormous gravitational field, or both.

All studies of Venus have failed to show any oblateness at all. It has no equatorial bulge and it seems to be just about a perfect sphere. Nor does it have an enormous gravitational field; in fact, the intensity of its gravitational field is a little less than that of Earth. Therefore, it must be rotating slowly.

The fact that Mariner II detected no magnetic field around Venus supported this conclusion, and the two findings together —neither oblateness nor a magnetic field—make it seem quite certain that Venus rotates slowly, much more slowly than Earth does.

BACKWARD ROTATION

But why should Venus rotate slowly?

To explain that, imagine two bodies revolving about each other—the Moon and Earth, for instance—with each rotating on its axis.

Earth's gravitational pull weakens with distance outward from Earth, so its pull on the side of the Moon facing away from us is a little less than the pull on the side of the Moon facing Earth. The difference in pull serves to stretch the Moon slightly, so that there is a very small bulge in its substance pointing directly toward the Earth, and another very small bulge pointing away from the Earth.

The Moon's gravitational pull on the Earth also decreases with distance, so it affects the far side of the Earth less than the near side, and this produces bulges on the Earth as well (see Figure 18). The effect on any object of a gravitational pull decreasing with distance is called a "tidal effect," because it is what produces the tides in Earth's ocean.

Figure 18
TIDAL EFFECTS

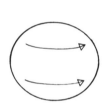

TIDAL EFFECT
OF EARTH ON MOON

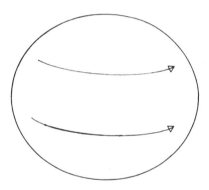

TIDAL EFFECT
OF MOON ON EARTH

As the Moon rotates about its axis, its rocky substances at any given point would rise and fall gradually as it passes across the side facing Earth, then rise and fall again as it passes across the side turned away from Earth; and would do so over and over again. This continuous rise and fall of the Moon's structure, twice each rotation, would produce a great deal of friction, and to overcome that friction there is a withdrawal of energy from the Moon's energy of rotation. The tidal effect, in other words, steadily slows the Moon's rotation and lengthens its period of rotation.

This slowing does not continue forever. Eventually the rotation is slowed to where it is exactly equal to the period of revolution about the larger body. Were that to happen, the Moon would rotate just quickly enough to keep one face to the Earth at all times. The tidal bulges would then remain where they were, one always facing the Earth, one always facing away from the Earth, and since the Moon would not be changing its position with respect to Earth, each bulge would remain at a certain point on the Moon's surface. The Moon's substance would not be rising and falling, because it would no longer be passing through that bulge, and there would therefore be no more slowing of rotation. The Moon would be "gravitationally locked" in position.

This has actually happened to the Moon. Its period of rotation and its period of revolution about the Earth are *both* equal to 655.7 hours, or 27.32 days. We always see the same face of the Moon. To see the other face, we must send out a probe equipped with a camera or human beings.

The Earth's rotation is also slowing because of the Moon's tidal effect, but the Earth is a much more massive body and has a much greater energy of rotation. Besides, the Moon's gravitational field is much smaller than the Earth's and has a correspondingly smaller slowing effect. The result is that although the Moon has slowed to the point of gravitational lock, the Earth has not, and won't for billions of years.

The Sun also produces tides on the Earth, and this contributes to the slowing effect on the Earth's rotation. The tidal effect, however, decreases with distance even more quickly than gravitation itself does. Therefore, even though the Sun is 26,500,000 times as massive as the Moon, its tidal effect on Earth (from its much greater distance) is only 0.44 that of the Moon.

Since Venus is considerably closer to the Sun than the Earth is, the Sun's tidal effect on Venus is considerably greater than its tidal effect on Earth—2.65 times as great, in fact. The Sun's tidal effect on Venus is even greater than the Moon's tidal effect on Earth—1.16 times as great.

For that reason, many astronomers felt that the Sun's tidal influence on Venus had been great enough to cause it to be gravitationally locked, with its period of rotation equal to that of the period of its revolution about the Sun. In that case, Venus's period of rotation would be 224.7 days.

What upset that seemingly straightforward deduction were the microwaves that, in the early 1960's, were sent to Venus and were reflected from it.

There are two things that should be mentioned in that connection. In the first place, although visible light is reflected from the cloud layer and cannot penetrate it, the much longer wavelengths of microwaves go through the clouds without trouble and are not reflected until they strike the solid ground beneath.

Because microwaves can do this, they give us an opportunity to study the solid surface of Venus, something we cannot do with ordinary light.

Secondly, the reflection of the microwaves is not changed from the original beam if the beam strikes a smooth and motionless surface. If the surface is moving, however, the wavelength is changed. From the amount of change, the speed of motion can be calculated. With microwaves, then, astronomers can tell what the period of rotation of Venus is.

The result was announced by Roland L. Carpenter and Richard M. Goldstein in 1962, even as Mariner II was on its way, and was a great surprise. It seemed that Venus was rotating so slowly that its period was roughly 250 days. That was distinctly *longer* than its period of revolution.

This was unique. There are bodies which are known to rotate in exactly the same period as that in which they revolve about a larger body. The Moon is an example. Other examples are the satellites of Mars and Jupiter. There are also many bodies known to rotate in periods that are smaller than their periods of revolution about a large body. Earth is an example of that. So are Mars, Jupiter, and Saturn.

Venus is the only body known to have a period of rotation longer than its period of revolution.

Could it be a mistake? Not at all! Careful examination of microwave reflections by the American physicist Irwin Ira Shapiro has refined the first measurements and shown that the period of rotation of Venus is 243.09 days.

This is a very slow rate of rotation. The circumference of Venus is 38,139 kilometers (23,699 miles), and if the planet turns in 243.09 days, then a point on the equator moves at only 6.54 kilometers (4.06 miles) per hour. This is only 1/255 of the equatorial speed of Earth, and only 2/5 of the equatorial speed of the slowly rotating Moon.

If you were standing on the surface of Venus and could see the stars (which, of course, you could not do because of the clouds), it would take 121.54 days for any one of them to march across the sky from rising to setting. If you walked briskly in the direction opposite to that in which Venus was rotating, you could cancel out the motion. (A human being can, after all, walk at a rate of four miles an hour, for a period of time at least.) A star that was overhead would therefore remain overhead for as long as you continued to walk briskly.

There is, however, something even more peculiar about Venus's rotation than the fact that its period is longer than that of its revolution about the Sun. To explain that, we must consider the motions of the planets.

Suppose we viewed the solar system from a point millions of kilometers above Earth's north pole and could observe the motions of the planets from there. We would see that all the planets move along their orbits in such a way as to revolve about the Sun in the same direction. All the planets circle the Sun in a counterclockwise direction (see Figure 19).

Figure 19
CLOCKWISE AND COUNTERCLOCKWISE

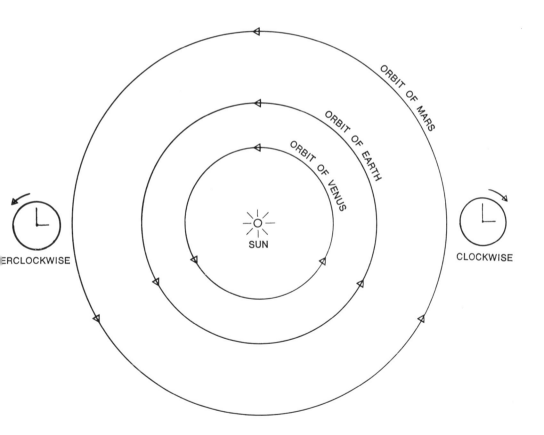

(Mind you, if we viewed the planetary motions from a position millions of kilometers above the Earth's *south* pole, they would all be in the clockwise direction. This does not result in confusion, though, because whenever astronomers talk about planetary motions, they always assume a point of view far above the north pole.)

That all the planets move in the same counterclockwise direction around the Sun is not likely to be a coincidence. This is one of the strong points in favor of the solar system's having started as a single swirling cloud of dust and gas. The planets all have the same direction of movement as the original cloud.

What is more, rotations are generally counterclockwise as well. The Earth turns about its axis in a counterclockwise direction. This is true also of Mars, Jupiter, Saturn, and the Moon. Then, too, satellites generally revolve about their planets in a counterclockwise direction.

Counterclockwise motion is so common in the solar system that it is taken as natural and is spoken of as "direct motion" or, sometimes, as "prograde motion," from Latin words meaning "step forward." It is taken, in fact, to be as normal a motion as the kind human beings use when they walk.

If a body of the solar system rotated or revolved in the clockwise direction, it would be called "retrograde motion," from Latin words meaning "step backward."

There are a few satellites of the outer planets, small and distant from those planets, that revolve in retrograde fashion, but it isn't common. In 1956 the American astronomer Robert S. Richardson suggested, however, that Venus might be rotating in retrograde fashion. He judged this from the wavelengths of light reflected at opposite edges of Venus. When Venus rotates, one side is approaching us and the other receding from us. From the side approaching us, light shortens its wavelength and from the side receding from us, light lengthens its wavelength. Considering the very slight differences in wavelength, Richardson decided that the "wrong" edge was receding and Venus was rotating backward.

No one paid much attention to Richardson's theory at the time, but when the microwave data were analyzed in 1962 there was no doubt about it. Not only was Venus rotating very slowly, but it was rotating in the retrograde direction.

The Earth rotates on its axis in such a way that its surface moves from west to east. Naturally, the Earth seems to us to be standing still and, instead, the stars seem to be moving in the opposite direction, from east to west. They rise in the east, move across the sky and set in the west. From the surfaces of the Moon, the Sun, Mars, and Jupiter, we would see the same east-to-west motion of the stars.

On Venus, however, the slow turning of the surface is from east to west. The stars, in Venus's sky, rise in the west, travel across the sky very slowly, and set in the east!

Let's take another look at the way a planet rotates. If a planet's rotation were exactly in the direction of its revolution about the Sun, its axis of rotation would be at right angles to the plane of revolution. If the plane of revolution is considered horizontal, the axis of rotation would be vertical.

This is not so in the case of Earth. Earth's axis is tipped by 23.45 degrees to the vertical (see Figure 20). This is Earth's "axial inclination."

Once the data concerning Venus's rotation was obtained, it was also possible to determine its axial tipping. That turned out to be 3.4 degrees, considerably less than that of Earth.

But wait! Suppose Earth's axis were tilted not 23.45 degrees, but 90 degrees or 180 degrees (see Figure 21). At 90 degrees, while the planet moved in its orbit about the Sun horizontally, the motion of the surface would be at right angles to that—up and down. At 180 degrees, the motion of the surface would be

Figure 20
AXIAL INCLINATION

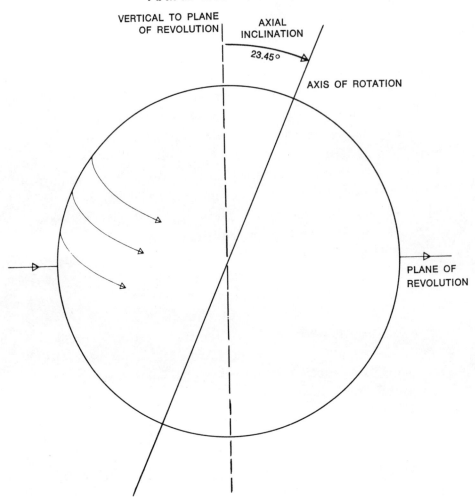

horizontal again, but the planet would now be moving in retrograde fashion.

In view of Venus's retrograde motion, we could say that its north and south poles were in opposite positions to those of Earth (see Figure 22). Since Venus's axis is at an angle of 3.4

Figure 21
EXTREME AXIAL INCLINATION

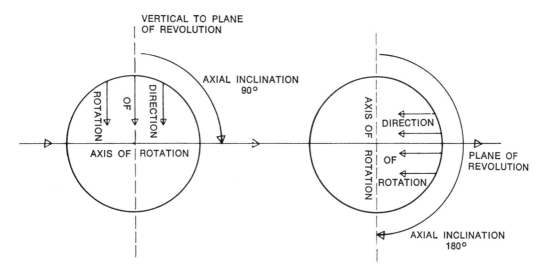

Figure 22
VENUS'S POLES

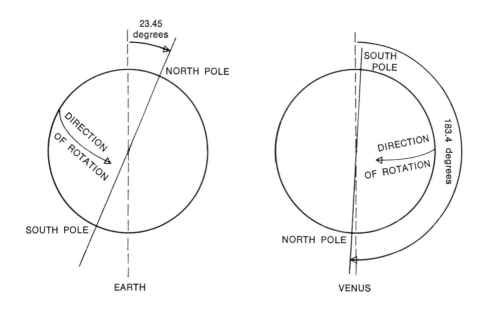

degrees to the vertical, we could say its actual axial inclination was 183.4 degrees.

The axial inclination of various planets is given in Table 25.

Table 25
AXIAL INCLINATION

	Axial Inclination (degrees)
Venus	183.4
Earth	23.45
Moon	6.7
Mars	24.0
Jupiter	3.1
Saturn	26.7
Uranus	97.9
Neptune	28.8

Notice that only Uranus, aside from Venus, has a really large axial inclination. Uranus's axial inclination of 97.9 degrees is roughly right-angled, so that it rotates up and down compared to its motion about the Sun. In fact, since the tip is slightly more than 90 degrees, Uranus's rotational motion leans toward the retrograde.

That is only a slight effect, however. For true retrograde motion, Venus stands alone.

5
The Details of Venus

How long is Venus's day? In other words, how long does it take the Sun to go from sunrise to sunset, then back to sunrise from the standpoint of someone standing on some particular spot on Venus's surface and watching. (Of course, no human being could endure the temperature of Venus's surface but we can pretend—or imagine some inanimate detecting device performing the task.)

Since Venus makes one complete turn on its axis in 243.09 Earth days, you might expect it to take 243.09 Earth days from sunrise to the next sunrise—or half of that, 121.55 Earth days, from sunrise to sunset. And since the rotation is retrograde, the Sun would rise in the west and set in the east.

This would be true if Venus were motionless with respect to the Sun—if it were not revolving about it. In that case, as Venus rotated east to west, it would seem to someone standing on Venus's surface that it was the Sun that was moving west to east (see Figure 23).

Venus, however, is also revolving about the Sun, and in the usual counterclockwise fashion. This revolution also produces

Figure 23
THE SUN AND VENUS'S ROTATION

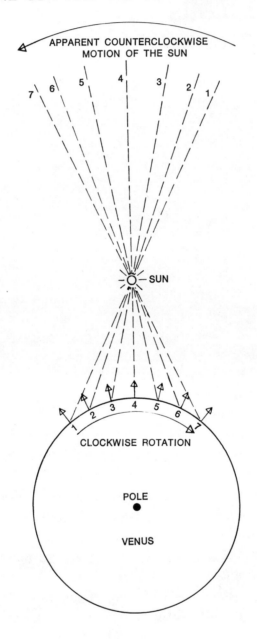

an apparent motion of the Sun. We can see what this is if we imagine that Venus is revolving about the Sun but not rotating on its axis at all. If that were so, then the conditions would be like that shown in Figure 24. As you see, the counter-clockwise revolution of Venus about the Sun causes the Sun to appear to move across the sky west to east, just as the clock-wise rotation of Venus about its axis does.

We can calculate how much each of Venus's motions, rotation and revolution, contributes to this apparent motion of the Sun.

Since Venus makes one complete turn about its axis in 243.09 Earth days, the Sun would appear to make one complete turn about the sky in 243.09 Earth days as a result of the rotation alone. There are 5,834.16 hours in 243.09 Earth days, so that in one hour the Sun would appear to drift west to east 1/5,834.16 of a complete circuit of the sky.

There are 360 degrees, or 21,600 minutes of arc, in a com-plete circle, so 1/5,834.16 of a complete circle is 21,600/5,834.16 or 3.702 minutes of arc. In one hour, then, the Sun drifts west to east in Venus's sky a distance of 3.702 minutes of arc, thanks to Venus's rotation.

But what about Venus's revolution? That, too, produces a west-to-east drift. Since Venus completes one revolution about the Sun in 224.70 Earth days, or 5,392.80 hours, Venus appears to make one west-to-east circuit of the sky in that same time. In one hour Venus drifts west to east 21,600/5,392.80 or 4.005 minutes of arc, as a result of revolution alone.

Altogether, counting the effects of both the rotation and the revolution, the Sun drifts west to east in Venus's sky a total of 3.702 + 4.005, or 7.707 minutes of arc each hour. If we divide that into the 21,600 minutes of arc that represent a complete circle of the sky, it turns out that the Sun appears to make a complete circle of the sky in 2,802.65 hours, or 116.78 Earth days.

Figure 24
THE SUN AND VENUS'S REVOLUTION

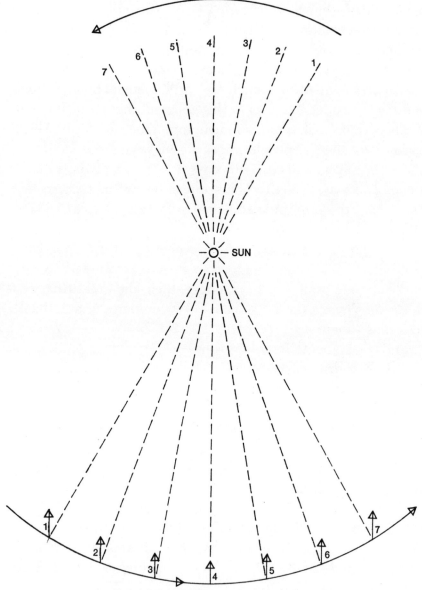

APPARENT COUNTERCLOCKWISE MOTION OF THE SUN

SUN

VENUS'S COUNTERCLOCKWISE REVOLUTION ABOUT THE SUN

Another way of putting it is that while Venus turns on its axis once in 243.09 Earth days relative to the stars—Venus's "sidereal day" (sigh-DEE-ree-al, from a Latin word meaning "star")—it turns on its axis once in 116.78 Earth days relative to the Sun (Venus's "solar day").

This sort of difference between the sidereal day and the solar day exists in all objects that revolve about the Sun and that also rotate on their own axes, but it is not generally so large a difference. The difference increases the more slowly the object rotates and the more rapidly it revolves. It is Venus's very slow rotation and fairly rapid revolution that make the two kinds of day so different in length in its case.

Where Earth is concerned, for instance, rotation on the axis is so fast that in the course of a complete day Earth moves only 1/365.2422 of the way about the Sun in its revolution, and that produces only a tiny shift of the Sun in the sky. Even so, it does introduce a small difference. Earth's solar day is exactly 24 hours long, but Earth's sidereal day—its rotation with respect to the stars—is only 23.934468 hours, 4 minutes shorter than the solar day.

Again, the Moon rotates on its axis and accompanies the Earth in its revolution about the Sun in such a way that the Moon's sidereal day is 27.295 Earth days long, while its solar day is 29.53 Earth days long.

The solar day is longer than the sidereal day in the case of the Earth and the Moon, but the solar day is shorter than the sidereal day in the case of Venus. That is because Earth and Moon rotate in the counterclockwise direction, while Venus rotates in the clockwise direction.

Next, imagine that Venus is at inferior conjunction and is just between the Earth and the Sun. The side toward us (call it Side A) is completely dark at that time. Since Venus's solar day is 116.78 Earth days long, Side A will be dark again 116.78

Earth days after inferior conjunction, again dark 233.56 Earth days afterward, again dark 350.34 Earth days afterward, again dark 467.12 Earth days afterward, and again dark 583.90 Earth days afterward.

But 583.9 Earth days after inferior conjunction, Venus is at inferior conjunction again. Venus, in other words, goes through exactly five solar days between inferior conjunctions, making just five complete turns with respect to the Sun. Each time Venus is at inferior conjunction, Side A—always the same side— is facing us.

Astronomers don't know why this is so. It would seem that Earth's gravitational field and, in particular, the tidal effects resulting from that field have caused Venus to lock into position with respect to us. It is because of this gravitational lock, perhaps, that Venus is forced to rotate in the retrograde direction at the very slow rate that it does.

The catch is that the Sun's tidal influence on Venus is about 16,500 times as strong as Earth's tidal influence even under conditions most favorable to Earth. Why, then, should Venus lock its rotation to Earth rather than to the Sun?

The curious fact that the synodic period of Venus is exactly five times its solar day is not satisfactorily explained. It is, as yet, one of Venus's mysteries.

VENUS'S ATMOSPHERE

From Earth, all we can see of Venus's atmosphere is its clouds.

Naturally, astronomers wanted to know more about the atmosphere than just the fact that it had a cloud layer. How dense was Venus's atmosphere? What gases made it up and in what proportion? How did its temperature vary with height, and so on?

To find out the answers, Venus probes continued to be sent toward the planet after the success of Mariner II. In 1967, for instance, the American probe Mariner V passed Venus at a distance of only 4,000 kilometers (2,500 miles) from the surface, and in February 1974 Mariner X skimmed by Venus, taking photographs.

Meanwhile, the Soviet Union sent a series of "Venera" probes toward Venus, the first two flying by, and those that followed plunging into the atmosphere and dropping to the surface by parachute. The Venera probes that entered the atmosphere sent back no information in the first few cases. Conditions on Venus were so much more extreme than expected that the probes were destroyed before they could do anything. The Soviet Union designed hardier and hardier probes that would endure higher temperatures and pressures for longer periods of time, and after that, information began to come back.

Taking advantage of the information from the Venera probes, the United States launched "Pioneer Venus" on May 20, 1978. It arrived on December 4, 1978, and went into a highly eccentric orbit about Venus. It passed very nearly over Venus's poles, circling the planet in 24 hours at a height ranging from 66,000 kilometers (41,000 miles) above the surface at its farthest point, down to 142 kilometers (88 miles) above its surface at its nearest.

Several probes left Pioneer Venus and entered Venus's atmosphere, confirming and extending the data earlier reported by the Soviet probes. Venus's surface temperature is 465° Celsius (870° Fahrenheit) and its atmosphere is about 90 times as dense as Earth's.

The main cloud layer on Venus is 3 to 4 kilometers (1.8 to 2.5 miles) thick and is centered at an altitude of 49 kilometers (30 miles) above the surface.

The liquid drops in the cloud layer include a considerable

quantity of sulfur, which gives Venus its yellowish appearance. Above the main layer of clouds are droplets of sulfuric acid. (The presence of sulfuric acid—a highly corrosive substance— had been suggested by several astronomers as long before as 1973.)

Below the cloud layer is a haze down to a height of 32 kilometers (20 miles) above the solid surface, and below that Venus's atmosphere seems completely clear. The lower atmosphere seems stable, without storms or weather changes—just incredibly steady heat everywhere.

Of the sunlight striking Venus, almost all is either reflected or absorbed by the clouds. Only 3 percent of the sunlight penetrates into the clear lower reaches of Venus's atmosphere, and perhaps 2.5 percent strikes the surface itself. (Compare this with Earth where, on the average, 30 percent of the incoming sunlight reaches the surface.)

Nevertheless, the sunlight reaching Venus in the first place is considerably more intense than that reaching Earth because the former is closer to the Sun. That means that the sunlight reaching Venus's surface is about 1/6 as bright as the sunlight reaching Earth's surface—despite Venus's thick and permanent cloud layer. Put another way, the sunlight reaching Venus's surface through its clouds is about 77,000 times as bright as the full Moon seen from Earth. Venus may experience a rather dim daylight, but there is no question that it is daylight.

The atmosphere on Venus (like that on Earth) consists almost entirely of two gases, but not the same two gases. On Venus it is carbon dioxide and nitrogen; on Earth it is oxygen and nitrogen (see Table 26).

Venus's surface area is only nine tenths that of Earth, a fact that decreases the total volume of the former's atmosphere, comparatively speaking. Even so, the great density of Venus's atmosphere gives it an enormous mass. Even though nitrogen

Table 26
ATMOSPHERE

| | Atmospheric Components (percentage) | | Atmospheric Mass (Earth's atmosphere = 1) | |
	Venus	Earth	Venus	Earth
Carbon Dioxide	96.6	—	78.5	—
Nitrogen	3.2	78.1	2.6	0.78
Oxygen	—	21.0	—	0.21

makes up only 3.2 percent of Venus's atmosphere, that nitrogen is 2.6 times as massive as Earth's total atmosphere.

Even the minor constituents of Venus's atmosphere are more important than they look. There may be only as little as 0.1 to 0.5 percent water vapor in Venus's atmosphere, but that is enough to make the mass of Venus's water vapor equal to that of the oxygen in Earth's atmosphere. The total water on Venus may be equal to 1/6000 that on Earth. That's enough water to cover New York State with a layer a mile high.

A gentle wind near Venus's surface (where there are only gentle winds) would, considering the greater mass of that wind, be sufficient to knock down earthly houses if any such could exist on Venus.

The atmosphere introduces several problems that astronomers are still puzzling over.

For one thing, Mariner X found that the upper reaches of the atmosphere whip around the planet in a clockwise direction —the same direction as that of the surface, but much more quickly. The upper-atmosphere winds circle the planet in nearly 4 Earth days, as compared to the 243 Earth days it takes the planet itself to make the circuit. Why the upper atmosphere speeds so and what keeps it going are not yet known.

For another, some of the trace components of Venus's atmosphere are present in surprising quantities.

Consider the gas argon. It is made up of inert atoms that do not combine with other atoms, and it should not have survived the early hot days of Earth. It should have leaked off into space. Other gaseous substances survived by temporarily combining with the molecules in Earth's rocky exterior, but argon could not have done this. It would be expected, then, that argon would form a very minor portion of the atmosphere, yet this is not so. Argon makes up a surprising 0.93 percent of Earth's atmosphere.

Most argon, however, is a variety known as argon-40, which is formed by the slow breakdown of a variety of potassium known as potassium-40. This breakdown has been taking place ever since Earth was formed and is still taking place today. Almost all the argon-40 now in the atmosphere appeared only after the Earth cooled to the point of being able to hold it.

There is, however, another variety of argon, argon-36, which is not formed by the breakdown of other atoms, and such of it as exists in the atmosphere has been there from the beginning. There is only one atom of argon-36 for every 300 atoms of argon-40, so argon-36 makes up only 0.003 percent of Earth's atmosphere. This seems as it should be.

Since Venus is closer to the Sun and must have been hotter than the Earth at every stage of its history, it should have lost its lighter atoms more rapidly than Earth did. It should therefore have even less argon-36 than Earth has. This apparently is not so. Venus's atmosphere seems to possess a hundred times as much argon-36 as Earth's.

Astronomers are still trying to puzzle that one out.

Finally, if Venus and Earth are so alike in size and general composition, how did they end up so different in atmosphere, temperature, and so on? Here astronomers have worked out a possible scenario.

Suppose that Venus and Earth started as worlds that were much alike, but with Venus closer to the Sun and, therefore, getting more heat and high-energy radiation. Both might have started with thin atmospheres containing carbon dioxide and nitrogen, and both might have had oceans of water. Because of the carbon dioxide in the atmosphere, both might have been warmer than Earth is now, with Venus somewhat the warmer of the two.

On Earth, the ocean was just cool enough for life to form early in its history. Eventually, primitive life forms developed ways of forming living tissue out of carbon dioxide and water, making use of the energy of sunlight to do so. That tied up some of the carbon dioxide of the atmosphere, replacing it with oxygen. That, in turn, reduced the greenhouse effect, lowering the temperature. The lower the temperature, the greater the amount of carbon dioxide that can dissolve in water. The atmosphere lost additional carbon dioxide, the ocean gained it, and the temperature continued to drop.

Life on Earth flourished as more carbon dioxide was available in the ocean, and a cycle was set up in which the atmosphere replaced carbon dioxide with oxygen at a steadily quickening rate and the temperature dropped to its present level, more or less.

On Venus, the ocean would have been warmer to begin with, and it would have evaporated to a greater extent so that there would be substantially more water vapor in Venus's atmosphere than in Earth's. Water vapor contributes to the greenhouse effect, so the temperature would have increased still further. The ocean may have become too warm for life just before life had a chance to form.

As Venus's temperature rose, without life forms to tie up the carbon dioxide, the warmer oceans released the carbon dioxide they could no longer hold in solution. This would cause the temperature to rise still further, even to the point where the carbonates (rocky substances that hold carbon dioxide in combination) in Venus's crust would release some of *their* carbon dioxide. (Nothing like that happened on Earth.)

The more carbon dioxide entered Venus's atmosphere from the ocean and the rocks, the higher the temperature rose and the more rapidly the ocean and the rocks gave off still more carbon dioxide. There was a "runaway greenhouse effect."

Eventually the oceans boiled, the water vapor in the upper atmosphere was broken down by the Sun's ultraviolet radiation, the hydrogen escaped, the oxygen formed sulfuric acid by combining with sulfur and water, and the Venus that exists today came into shape.

The actual temperature of Venus seemed, however, to be too high to be accounted for by even the vast quantity of carbon dioxide in the atmosphere. Recent findings, however, show that more sunlight reached the surface than had been expected, and so some of the minor components discovered in Venus's atmosphere could contribute importantly to the greenhouse effect. The sulfur in the clouds and a 0.02 percent content of the gas sulfur dioxide in Venus's -atmosphere both trap significant amounts of heat.

The chief uncertainty is whether Venus could possibly have had an ocean as large as Earth's to begin with. There is some difficulty in accounting for the loss of all that water vapor.

As we look over the scenario, then, remember that Venus and Earth might well have been very much alike to begin with, except that Venus was a little warmer. That difference in temperature was enough to make them take different roads, with the result that the two planets are the radically different worlds they are today.

This may show how small a change it would require to make Earth uninhabitable in its turn, and may explain why we should be particularly careful about burning so much coal and oil which are, right now, adding many tons of carbon dioxide to Earth's atmosphere each year.

VENUS'S SURFACE

What about the solid surface of Venus beneath the obscuring cloud layer?

On October 21, 1975, Venera 9 landed on Venus's surface. It was not only sturdy enough to function for a period of time, but it was equipped to take photographs and relay them to Earth.

There were surprises at once. Scientists had been so sure that there would be little light under the thick clouds that Venera 9 was fitted with a lighting system by which the photographs could be taken. The lighting system wasn't needed. The bright Sun of Venus was able to get enough light through the clouds to light up the landscape to the level of a cloudy day on Earth.

Apparently, Venera 9 landed on a mountain that was well over a mile above the average level of Venus's surface. (You can't speak of sea level, since there is no sea.) The temperature was therefore a little low, only 448° Celsius (838° Fahrenheit), and the pressure was likewise a little low, only 85 times Earth's atmosphere. That helped the probe to continue functioning for 53 minutes before it ceased working. The pictures showed a rough landscape with nearby boulders of about 1/3 meter (a foot or so) across.

Four days later, Venera 10 landed about 1400 miles from Venera 9 and closer to Venus's equator. It took pictures for 65 minutes and showed rocks that were flat and wide. In both cases the rocks had sharp edges, which showed that not much erosion had taken place.

Surface winds were recorded that weren't very fast, only a little over 11 kilometers (7 miles) an hour. Since the atmosphere of Venus is so dense, however, such winds would have the energy of earthly winds blowing at 105 kilometers (65 miles) an hour. The "gentle" wind is just about equivalent to a hurricane on Earth.

While such probes can study tiny patches of Venus's surface in great detail for an hour or so, more can be found out by radar.

Beams of microwaves, striking Venus and being reflected, can yield information about the surface just as reflected light would. Light, of course, would be better because we can analyze the light directly by eye, whereas for radar we must use complicated instruments. Then, too, microwaves have much longer wavelengths than light does, and that gives us a fuzzier picture, one that misses the fine details. It is as if a long-legged human being stepped over irregularities without noticing them, while an ant, having to clamber over those same irregularities, would be aware of each in detail.

Nevertheless, microwaves are long enough to go through the clouds on Venus both coming and going, while light waves can't. It is better to have a fuzzy picture than none at all.

Radar studies could be made from Earth itself, but here we run into the peculiarity that the same side of Venus always presents itself to Earth at inferior conjunction when Venus is nearest and can be studied from Earth in greatest detail. This means that Earth-based radar studies show us only half of Venus and are relatively insensitive to the equatorial region of even that half.

Even so, Earth-based radar studies showed the presence of

a couple of mountain ranges, a possible large volcano, and a giant canyon.

Radar studies came into their own, however, with "Pioneer Venus." This probe approached Venus more closely on its daily swoops in toward the planet and reached different points from day to day, so that we now have a radar map of almost the entire surface of the planet.

From this map, it would appear that Venus's crust is quite different from that of Earth. Earth's crust is relatively thin and is broken into half a dozen large plates and a number of smaller ones. These plates move very slowly relative to each other, crushing together, or slipping one under another, or pulling apart. These movements, called "plate tectonics," produce earthquakes and volcanoes at the joints, build mountain ranges where the plates crush together, ocean deeps where one shoves under another, rifts where they pull apart.

Venus, on the other hand, may have gone through such a period early in its history, but it seems that at present its crust is more or less all one piece. Most of Venus's surface seems to be the kind we associate with continents rather than sea bottom. We might assume that whereas on Earth we have a vast sea bottom making up 7/10 of the planet, with the continents placed within it like large islands, Venus has a huge super-continent that covers about 5/6 of the total surface, with small regions of lowland making up the remaining sixth.

It may be, then, that Venus, when it was originally cool, did not have an ocean as Earth had, but, at best, inland seas. It was perhaps the lack of water that helped encourage the runaway greenhouse effect. It made it easier for the seas to boil, and it makes it easier to explain the action of ultraviolet light in removing enough of the water in the upper atmosphere as to leave so low a water content on Venus today.

The supercontinent that covers Venus seems to be rather level, with some indications of craters, but no indication of large quantities of these. The thick atmosphere may have eroded them away. There are, however, raised portions on the supercontinent, two of them being of huge size.

In what on Earth would be the arctic region, on Venus there is a large plateau which is named "Ishtar Terra," or Land of Ishtar. (Ishtar is the Babylonian goddess who is the equivalent of the Roman goddess Venus.) Ishtar Terra is about as large as the United States.

The western portion of Ishtar Terra is relatively flat and is about 3.3 kilometers (2 miles) above the ordinary level of Venus. On the eastern portion is the mountain range "Maxwell Montes," which was one of the features of Venus detected from Earth. The highest peaks in Maxwell Montes are as much as 11.8 kilometers (7.3 miles) above the levels outside the plateau. This is 3 kilometers (1.9 miles) higher than Mount Everest on Earth.

In the equatorial region of Venus there is another and even larger plateau called "Aphrodite Terra" (Aphrodite is the Greek equivalent of the Roman Venus). It is 9,600 kilometers (6,000 miles) wide and has some peaks that are 8 kilometers (5 miles) high, or roughly as high as the Himalayan range on Earth.

From the eastern end of Aphrodite Terra, there is a group of great canyons that extend for some 5,000 kilometers (3,100 miles). They are extremely deep, some splitting the crust to a depth of 2.9 kilometers (1.8 miles) below the average level of Venus's surface.

It is hard to tell whether any of the mountains of Venus are actually volcanoes. "Beta Regio," which was first detected from Earth, seems, on the Pioneer Venus radar map, to possess two mounts, "Rhea Mons" and "Theia Mons," each about 4 kilometers (2.5 miles) high. Rhea Mons may spread across an area

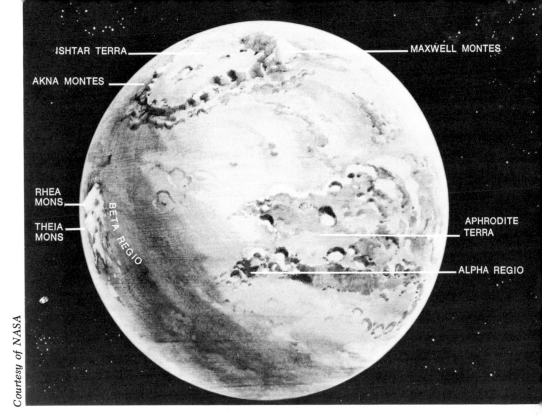

ISHTAR TERRA — MAXWELL MONTES

AKNA MONTES

RHEA MONS

BETA REGIO

THEIA MONS

APHRODITE TERRA

ALPHA REGIO

Contour map of Venus.

as large as New Mexico, and may be larger than Olympus Mons on Mars.

These plateaus, mountain ranges, volcanoes, and canyons may all have been formed in the early days of Venus when plate tectonics was still in process—but they all make up only 5 percent of Venus's surface.

Naturally, astronomers hope they will get much clearer pictures of Venus's surface in years to come, but only thirty years ago it might well have seemed that no one would *ever* get a chance to see beneath Venus's clouds.

VELIKOVSKY

Our new knowledge of Venus aids astronomers in their consideration of the theories of Immanuel Velikovsky, a Russian-American psychiatrist.

In 1950 Velikovsky published *Worlds in Collision,* in which he suggested that Venus had been ejected from Jupiter and had taken up its present orbit about 1500 B.C.

In traveling from Jupiter to its present position, it had passed near the Earth and had produced a series of dire events memorialized in the Bible as the ten plagues that visited Egypt in the time of Moses. Some forty years later, it also momentarily stopped Earth's rotation (recounted in the Bible in connection with Joshua's command that the Sun and Moon stand still) and then restarted it again. In still later years, Venus had a near brush with Mars.

All this is so far out of line with astronomical conclusions based on centuries of observation and scientific thought that very few astronomers take it seriously for even a moment.

For instance, Jupiter's structure is now known to be widely different from that of Venus, so that the possibility of a rocky and metallic world like Venus being ejected by a gaseous and liquid hydrogen world like Jupiter is just about zero. To be sure, Jupiter *may* have a rocky, metallic core, but there would seem to be no reasonable way in which a piece of that core would explode outward through all the layers that lie overhead.

Then, too, the path of Venus must have taken up an elongated orbit about the Sun for it to move from Jupiter to a close approach to Earth. It must then have settled down into the nearly circular orbit that Venus has today. This simply cannot be explained by any of the rules of celestial mechanics that have proved so useful to astronomers since the time of Kepler.

Again, if Venus had stopped the rotation of the Earth, all the energy of the rotation would have been converted into heat and would have boiled the oceans. To have Venus then restart the rotation of the Earth and restore it to what it was (as nearly as we can tell), to the exact second, passes the bounds of belief.

These are just a few of the many, many follies of the Velikovskian notions, which would not have been worth a glance except that they pretended to explain some of the "miracles" of the Bible and therefore won a large following. There are many "Velikovskians" (few of whom know any astronomy at all) who are convinced that Velikovsky's notions are right and that astronomers are engaged in a conspiracy to suppress the truth of his theories.

To be sure, Velikovsky made some predictions that seemed to be close to what astronomers eventually discovered to be so, and the Velikovskians make the most of that.

For instance, Velikovsky stated that since Venus was formed from Jupiter's interior which must be very hot, Venus itself must be very hot. He said this in 1950, when astronomers believed that Venus's temperature, while warmer than Earth's, might not be very much warmer.

Had astronomers known in 1950 how dense Venus's atmosphere was and how rich in carbon dioxide it was, they would have realized then that Venus would have to be hot—but it was over a decade before that became clear and, meanwhile, Velikovsky had beaten them to the punch.

What's more, Venus proved to be *so* hot that astronomers had trouble accounting for so high a temperature even by supposing a runaway greenhouse effect, whereas Velikovsky had said all along that Venus was nearly, or quite, red-hot.

Could it be that Velikovsky was right in supposing Venus to have emerged from Jupiter just 3500 years ago?

Well, if Venus's heat is merely inherited from Jupiter, it should be cooling off, and Velikovsky's notion was that it had been cooling since it was born and is still cooling off. On the other hand, if astronomers are right in supposing Venus to be hot because of a greenhouse effect (even if they are not sure of all the details as yet), then the temperature should have been holding steady for perhaps billions of years and should be holding steady now.

The best measurements Pioneer Venus could make indicate that Venus gives off just as much heat at night as it gains from the Sun by day. Its temperature is steady and it is not cooling off, so in this respect the astronomers seem to be right and the Velikovskians wrong.

The recent discovery that argon-36 is unexpectedly high in Venus's atmosphere could be interpreted to mean that it is a young world. It picked up the argon-36 in Jupiter's atmosphere as it passed through and has not yet had time to lose it. However, if Venus emerged from Jupiter white-hot (as it would have had to be considering that Jupiter's interior is at a temperature of 50,000° Celsius or 90,000° Fahrenheit), it would surely have lost any argon-36 it had picked up—if it could have picked up any in the first place. Undoubtedly, whatever the explanation of argon-36 in Venus's atmosphere, it will turn out to be something that doesn't involve anything as outlandish as Velikovsky's notions.

Then, too, the plateaus and mountains that have now been discovered on Venus's surface give every sign of having been in place for enormous lengths of time and not for merely 3500 years. Venus's crust must be very rigid to stand up under the weight of those large plateaus, and it could not have been molten as recently as Moses's time when it was supposed to have come whizzing, white-hot, out of Jupiter.

In addition, the Velikovskians, while talking interminably about those guesses that turn out to have been right, tend to ignore guesses that turn out to have been wrong.

In order to explain the manna that fed the Israelites in the desert, Velikovsky insisted that the atmosphere of Venus was hydrocarbon in nature. At one point some years ago, it appeared that methane (a simple hydrocarbon) had been detected in Venus's atmosphere. The Velikovskians leaped at that, and when the report turned out to be a mistake, they wouldn't let go. They kept on saying that there were hydrocarbons in Venus's atmosphere.

But there aren't. The only sizable components of Venus's atmosphere are carbon dioxide and nitrogen. Anything else is present in relatively small quantities. That is quite certain.

There seems to be no question that Velikovsky's theories, however interesting and dramatic they might be, are quite, quite wrong. There will no doubt continue to be Velikovskians who, for a variety of emotional reasons, will cling to those theories, but if people insist on being foolish there is no way of stopping them.

The cratered surface of the planet Mercury as photographed by Mariner X.

6

Mercury

ANOTHER EVENING STAR

Venus is closer to the Sun than Earth is, but it sets no record in that regard. There are other objects that are closer to the Sun, all or part of the time, than Venus is. The rest of the book will be devoted to those nearer neighbors of the Sun.

Since ancient times it has been known that Venus is not the only evening and morning star. There is another planet that moves through the sky as Venus does—but, in a manner of speaking, even more so.

At times this other planet appears to the west of the Sun, rising in the morning before the Sun and then vanishing in the glare of dawn. At times it appears to the east of the Sun, shining in the gathering twilight until it sets in its turn while night is yet young.

Like Venus, this second evening star and morning star was considered two separate bodies when it was first noticed. The Egyptians called it "Set" and "Horus" after two of their gods. The Greeks called it "Apollo" when it was a morning star, and "Hermes" when it was an evening star.

Apollo was a good name for the morning star, for Apollo was

the god of the Sun and he was pictured as driving the blazing chariot of the Sun. When Apollo rose as the morning star (usually later than Venus did) it seemed to be the driver indeed, with the blazing chariot coming very soon after.

It was eventually noticed, however, that when the evening star was present the morning star never was, and vice versa, and that what seemed like two planets was really but one. Like Venus, this second planet moved from one side of the Sun to the other, back and forth, and the Greeks came to apply the name Hermes to both the morning star and the evening star.

Later, the Romans called the planet "Mercurius" after a god they considered the equivalent of the Greek Hermes. In English, the name became Mercury and that is what we call the planet today.

Mercury is never as bright as Venus is at its brightest, and is never as far from the Sun as Venus is at its farthest. Mercury is, in fact, never more than 28 degrees from the Sun (see Figure 25), never sets more than two hours after sunset when it is an evening star, or rises more than two hours before sunrise when it is a morning star.

Mercury, then, is dimmer than Venus, usually; lower in the sky than Venus, usually; and closer to the Sun than Venus, usually. Mercury is almost never seen when the sky is fully dark. As a result, Mercury is not easy to observe, and people who live in cities hardly ever get a chance to see it with the unaided eye. The great astronomer Copernicus is supposed never to have seen it.

Enough people did see it, however, for its motions against the stars to be noted. One thing about those motions was discovered quickly. There were times when it shifted position against the stars more quickly than Venus did and more quickly than any of the heavenly bodies but the Moon.

That is why the ancient Greeks named it Hermes, because

Figure 25
MAXIMUM ELONGATION OF MERCURY

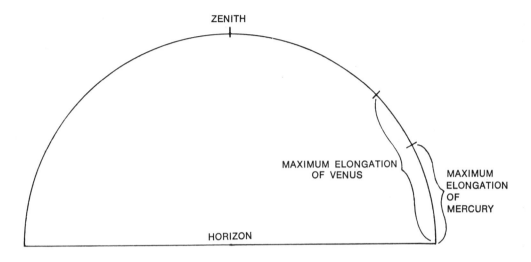

he was the messenger of the gods, who traveled so quickly on his errands that he was usually pictured with wings on his helmet and on his sandals. (The same applies to the Roman god, Mercury.)

When the ancients matched the seven known metals to the seven known planets, quicksilver was matched to Mercury. Quicksilver is the only liquid metal, and its name makes use of the meaning of "quick" as "alive" (as in "the quick and the dead"). Though quicksilver looked like silver, it was not a solid lump of matter, inert and dead. It easily split into drops that sped this way and that, almost as though it were alive.

Its ability to speed away made it natural to associate it with the speedy planet, Mercury, and in time quicksilver lost its own name and became known by the planet's name exclusively. We call that liquid metal "mercury" today.

Because the planet Mercury shifts position against the stars so quickly, it was thought to be closer to Earth than Venus is, closer to Earth than any body but the Moon. The ancient Greeks listed the planets in order of increasing distance from the Earth as: Moon, Mercury, Venus, Sun, Mars, Jupiter, and Saturn.

Once it was discovered that the planets circled the Sun and not Earth, the order outward from the Sun became: Mercury, Venus, Earth-Moon, Mars, Jupiter, and Saturn.

Mercury, as it moves through the sky, behaves much as Venus does. Mercury can pass between the Earth and the Sun and be at inferior conjunction. It can pass behind the Sun and be at superior conjunction. It is because its orbit about the Sun is smaller than that of Venus that Mercury's maximum elongation is less than that of Venus (see Figure 26).

Mercury shows phases exactly as Venus does. Mercury is in the full phase when it is at superior conjunction and farthest from us, and in the new phase when it is at inferior conjunction and nearest to us. At maximum brightness, it is a thick crescent.

THE DISTANCE AND SIZE OF MERCURY

At its brighest, Mercury has a magnitude of −1.2. It is then only 5/6 as bright as Sirius, the brightest star, and only about 1/17 as bright as Venus at its brightest.

Why?

To begin with, there is the question of distance. If you look at Figure 26, you will see that sometimes Mercury can be closer to us than Venus is, and sometimes the reverse is true. When both planets are at superior conjunction, Mercury is closer to us than Venus is; when both planets are at inferior conjunction, Venus is closer to us than Mercury is.

Figure 26
ORBITS OF MERCURY AND VENUS

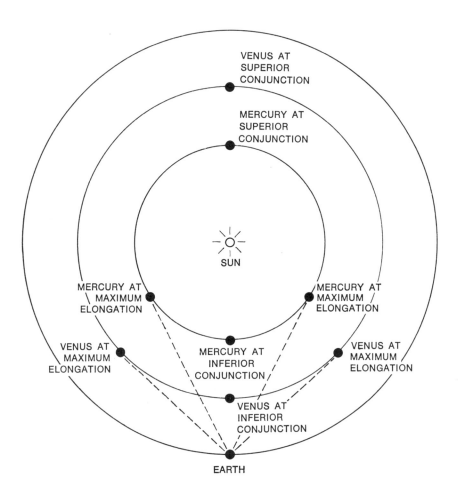

We can put some figures to this. Table 27 gives the average distance of Mercury, Venus, and Earth from the Sun. (*Average* distance is given because the planets do not stay the same distance from the Sun. Each has a perihelion and an aphelion and,

Table 27
DISTANCE OF MERCURY FROM THE SUN

| | Average Distance | | |
	Kilometers	Miles	Earth = 1
Mercury	57,900,000	36,000,000	0.387
Venus	108,200,000	67,200,000	0.723
Earth	149,600,000	92,960,000	1.000

as we shall see, this is particularly important in the case of Mercury.)

Given these figures, it is possible to work out the average distance of Venus from Earth and of Mercury from Earth at either inferior conjunction or superior conjunction (see Table 28). As you see, when both planets are at inferior conjunction,

Table 28
AVERAGE DISTANCE AT CONJUNCTION

| | Average Distance from Earth | | | |
| | At Inferior Conjunction | | At Superior Conjunction | |
	Kilometers	Miles	Kilometers	Miles
Mercury	91,700,000	57,000,000	207,500,000	129,000,000
Venus	41,400,000	25,800,000	257,800,000	160,200,000

Mercury is (on the average) some 2.2 times as far away from us as Venus is.

To be sure, neither inferior conjunction nor superior conjunction is important as far as the brightness is concerned, since in either case the planet is too close to the Sun to be seen. Each planet is at its brightest when it is moving out toward maximum elongation and some little time before it reaches it. At this time, Mercury tends to be about 1 1/3 times as far from us as Venus is.

The greater distance is in itself enough to cause us to expect Mercury to be dimmer than Venus. If distance were the only difference involved, however, one would expect Venus to be about 1.8 times as bright as Mercury and not 17 times as bright, as it is. There must be other factors involved as well.

What about Mercury's size? Mercury's apparent diameter in the sky is consistently smaller than that of Venus (see Table 29 and Figure 27). It is not surprising that Venus should ap-

Table 29
APPARENT DIAMETER OF MERCURY

	Apparent Diameter	
	At Inferior Conjunction (seconds of arc)	At Superior Conjunction (seconds of arc)
Mercury	12.7	4.7
Venus	61.0	9.9

pear larger than Mercury at inferior conjunction, since Venus is then closer to us than Mercury is, though Venus is larger in appearance than the difference in distance gives us a right to expect.

The real giveaway is the fact that Mercury is *closer* to Earth when it is at superior conjunction than Venus is under that same condition, yet even so, Mercury's apparent diameter at superior conjunction is a trifle less than half of Venus's.

Figure 27
APPARENT SIZE OF MERCURY

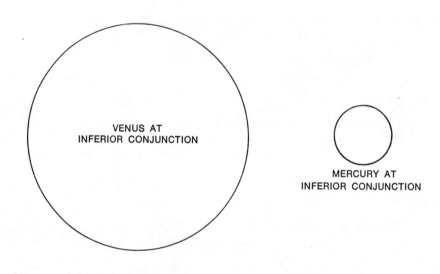

VENUS AT
INFERIOR CONJUNCTION

MERCURY AT
INFERIOR CONJUNCTION

VENUS AT
SUPERIOR CONJUNCTION

MERCURY AT
SUPERIOR CONJUNCTION

To account for this, Mercury must be a smaller body than Venus is in actuality and not just in appearance—and so it is. From the apparent size and actual distance, astronomers can calculate the actual diameter of Mercury as compared to Venus and Earth (see Table 30).

Table 30
DIAMETER OF MERCURY

| | Diameter | | |
	Kilometers	Miles	Earth = 1
Mercury	4,850	3,014	0.38
Venus	12,112	7,526	0.95
Earth	12,756	7,927	1.00

Mercury is considerably smaller than either Venus or Earth, with roughly only 1/3 the diameter of the larger planets.

To be sure, Mercury is not altogether insignificant. It is substantially larger than our Moon, for instance. Nevertheless, Jupiter's largest satellite, Ganymede, and Saturn's largest satellite, Titan, are each considerably larger than the planet Mercury (see Table 31 and Figure 28).

Table 31
MERCURY AND THE SATELLITES

	Diameter (Moon = 1)	Surface Area (Moon = 1)	Volume (Moon = 1)
Moon	1.00	1.00	1.00
Mercury	1.40	1.95	2.71
Ganymede	1.50	2.25	3.38
Titan	1.67	2.79	4.65
Earth	3.68	13.54	49.82

The surface area of Mercury is about 74,600,000 square kilometers (28,800,000 square miles), or that of Asia and Africa

Figure 28
MERCURY AND THE SATELLITES

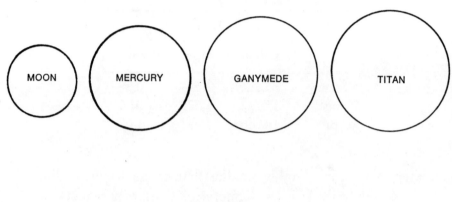

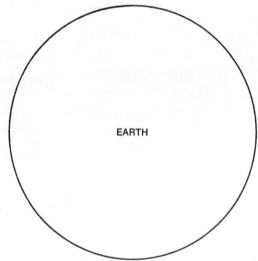

combined. If Mercury were hollow, nearly three bodies the size of our Moon could be packed into it. On the other hand, Titan could hold nearly two bodies the size of Mercury, and if Earth were hollow it would take 18 bodies the size of Mercury to fill it.

If we took into account the distance and size of Venus and

Mercury, we would expect Venus to be about 11 times as bright as Mercury when both are at their brightest, but Venus is, in actuality, 17 times as bright. Something is still left out.

Mercury's average distance from the Sun is a little over half that of Venus. The amount of light any object receives from the Sun increases as the distance decreases, and the increase is a rapid one that is related to the square of the distance. In other words, if Mercury were twice as close as Venus it would get 2 × 2 or four times the light; if it were three times as close it would get 3 × 3 or nine times the light, and so on. In actual fact, Mercury gets 3 1/2 times as much light from the Sun, on the average, as Venus does.

That should make Mercury seem brighter than one would expect from its size and distance only, instead of dimmer— but who says that Mercury reflects all the light it gets? Venus reflects 61 percent of all the light it receives from the Sun. Because of its high albedo, it gleams brilliantly. Mercury, on the other hand, reflects only 6 percent of the light it receives.

Why does Mercury reflect so little of the light that falls on it?

Because unlike Venus (but like our Moon), Mercury has no significant atmosphere. It is a world with a surface of dark, bare rock, and this absorbs most of the light that falls on it.

There are the reasons why Mercury is so much dimmer than Venus: Mercury is a smaller world than Venus is; it is at a greater distance from Earth than Venus is; and, unlike Venus, it has no atmosphere.

THE ORBIT OF MERCURY

Since Mercury's orbit is smaller than that of Venus, we would expect Mercury to complete a revolution about the Sun in less time than it takes Venus to do so, and certainly in less time than it takes Earth to do so. That turns out to be indeed true (see Table 32).

Table 32
PERIOD OF REVOLUTION OF MERCURY

	Period of Revolution		
	Years	Months	Days
Mercury	0.241	2.89	88.0
Venus	0.615	7.38	224.7
Earth	1.000	12.00	365.24

An 88-day period of revolution for Mercury is shorter, perhaps, than we might expect from the length of its orbit. Mercury's orbit is about 2/5 as long as Earth's orbit, yet it takes Mercury considerably less than 2/5 of a year to race around it. Not only does it have a shorter orbit, but it races along that orbit more quickly. But then, it is closer to the Sun and therefore is subject to a greater intensity of the Sun's gravitational field. That accounts for its greater orbital speed (see Table 33).

How often does Mercury reach inferior conjunction? If Earth were standing still, then Mercury would complete its revolution in 88 days and, with that, move between the Earth and Sun again. While Mercury made its revolution, however, Earth would, in actual fact, be moving also. By the time Mercury made one complete turn about its orbit, Earth would have moved on and completed nearly 1/4 of its turn. Mercury must pursue it, but because its period of revolution is shorter than that of Venus, Earth gains less distance. Then, because Mercury moves more quickly than Venus does, Mercury catches

Table 33
ORBITAL SPEED OF MERCURY

	Orbital Speed	
	Kilometers per Second	Miles per Second
Mercury	47.89	29.76
Venus	35.03	21.77
Earth	29.79	18.51

up to Earth's lesser gain a lot sooner than Venus would. The time lapse between inferior conjunctions (the synodic period) is thus much shorter for Mercury than for Venus (see Table 34).

Table 34
SYNODIC PERIOD OF MERCURY

	Synodic Period		
	Years	Months	Days
Mercury	0.32	3.81	115.88
Venus	1.60	19.2	583.92

If Mercury's orbit lay in the same plane as Earth's, then every time Mercury was at inferior conjunction it would be exactly between Earth and Sun and there would be a transit—it would be seen to cross the face of the Sun as a tiny black orb. This doesn't happen every time because the plane of Mercury's orbit is slightly inclined to that of Earth's orbit, just as is true of Venus (see Table 35 and Figure 29).

Table 35
ORBITAL INCLINATION OF MERCURY

	Orbital Inclination (Degrees)
Mercury	7.00
Venus	3.39

Figure 29
ORBITAL INCLINATION OF MERCURY

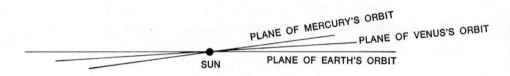

You might think that its greater inclination would cause Mercury to miss passing in front of the Sun at inferior conjunction to a greater extent than Venus would, and that therefore transits of Mercury would be even rarer than of Venus.

This is not so. In the first place, because Mercury moves so quickly and has such a short synodic period, there are five times as many inferior conjunctions of Mercury in a given period than there are inferior conjunctions of Venus. Secondly, the closer a planet is to the Sun, the greater the orbital inclination required to make it miss the Sun as viewed from the Earth. Mercury is sufficiently close to the Sun that even its greater orbital inclination doesn't cause it to miss as many times as Venus does.

There are, on the average, 13 transits of Mercury every century, but they don't come regularly. The interval between transits can be 3, 7, 10, or 13 years.

Transits of Mercury only take place when the Earth happens to be at that point in its orbit which crosses the plane of Mercury's orbit (and then only when Mercury happens to be near that point, too, at the time.) Earth crosses the plane of Mercury's orbit in May and then again half a year later in November. Transits of Mercury always take place, therefore, either in May or November.

Kepler was the first to work out the planetary orbits of the solar system well enough to be able to predict a transit of Mercury. He predicted one for November 7, 1631, and on that day a French scientist, Pierre Gassendi (gah-sahn-DEE), was watching. Gassendi caught the transit, even though it came five hours ahead of Kepler's calculation. Since then, every transit of Mercury has been caught by one astronomer or another.

ORBITAL ECCENTRICITY OF MERCURY

Venus and Earth have such low orbital eccentricities that the ellipses of their orbits are so rounded as to look like circles to the casual observer. Not so in the case of Mercury (see Table 36).

Table 36
ORBITAL ECCENTRICITY OF MERCURY

	Orbital Eccentricity
Mercury	0.2056
Venus	0.0068
Earth	0.0167

Mercury's large orbital eccentricity means that its orbit is an ellipse that is flattened enough to look like an ellipse. This also means that the two foci of the ellipse are separated by quite a bit (see Table 37 and Figure 30).

Table 37
SEPARATION OF FOCI

	Separation of Foci	
	Kilometers	Miles
Mercury's orbit	24,000,000	15,000,000
Venus's orbit	1,400,000	900,000
Earth's orbit	5,100,000	3,200,000

Figure 30
ORBIT OF MERCURY

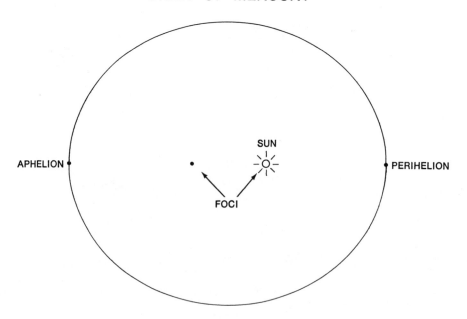

Since the Sun is always at one of the foci of an elliptical orbit, it is considerably closer to one end of Mercury's orbit than to the other. Consequently, Mercury's distance from the Sun changes a good deal as it travels along its orbit. At perihelion, the distance is considerably smaller than average, and considerably larger than average at aphelion (see Table 38).

Table 38
CHANGING DISTANCE OF MERCURY
FROM THE SUN

	Distance from the Sun	
	Kilometers	Miles
Mercury at Perihelion	45,900,000	28,500,000
Mercury at Aphelion	70,000,000	43,500,000

This unevenness in the distance of Mercury from the Sun has its effects. For instance, the farther Mercury is from the Sun, the closer it can approach Venus and Earth, if those planets happen to be on the same side of the Sun as Mercury is (see Table 39).

As you see, although Venus is Earth's closest neighbor, Earth is not Venus's closest neighbor. Venus's closest approach to Earth is 41,400,000 kilometers (25,800,000 miles), but when Mercury is at aphelion, Venus can be 3,200,000 kilometers (2,000,000 miles) closer to Mercury than it is to Earth.

It happens that Mercury is at aphelion in a point of its orbit that points nearly toward the aphelion portion of Earth's orbit, when Earth is also farther than usual from the Sun. The same is true of the two perihelion points for the two planets. Since both Earth and Mercury are relatively far from or relatively

Table 39
DISTANCE OF MERCURY FROM VENUS AND EARTH

	Minimum Distance from Venus		Minimum Distance from Earth	
	Kilometers	Miles	Kilometers	Miles
Mercury at Perihelion	62,300,000	38,700,000	101,200,000	62,900,000
Mercury at Aphelion	38,200,000	23,700,000	82,200,000	51,100,000

near to the Sun at the same time, that cuts down the change in distance between them a bit.

The perihelion and aphelion positions don't stay in the same spot in space as the planets whirl about the Sun. Very slowly, the perihelion and aphelion shift along the orbit in the case of both Mercury and Earth, and at different speeds. There will come a time, many years from now, when the aphelion portion of Mercury's orbit will point toward the perihelion portion of Earth's orbit. Mercury will then be farthest from the Sun at a time when Earth is nearest to the Sun, and if Mercury is at inferior conjunction at that time, it will be only 77,000,000 kilometers (47,900,000 miles) from Earth. That is as close as it will ever get, and it will still be twice as far from us as Venus can be.

The ellipticity of Mercury's orbit affects how easily it may be seen. Mercury's maximum elongation of 28 degrees takes place when it happens to be at or near aphelion. That is when it can be seen as long as two hours after sunset or before sunrise.

If Mercury is at perihelion at the time of maximum elongation, it is only 18 degrees from the Sun, setting only about an hour after the Sun if it is an evening star, or rising about an hour before the Sun if it is a morning star. Mercury is hard enough to see when it is at aphelion at the time of maximum elongation; the difficulties are multiplied when it is at perihelion.

When Mercury reaches inferior conjunction near one of its two points of crossing Earth's orbital plane, it happens that the May position is close to Mercury's aphelion and the November position close to its perihelion. The closer Mercury is to the Sun, the less effect its orbital inclination has in making it miss the Sun. This means that a transit is more likely to occur in November than in May, and, in fact, there are twice as many transits in November as in May.

Another effect on Mercury of its changing distance from the Sun is the change in its orbital speed. The closer it is to the Sun, the more intense is the Sun's gravitational effect on it and the faster it moves (see Table 40).

Table 40
CHANGING ORBITAL SPEED OF MERCURY

	Orbital Speed	
	Kilometers per Second	Miles per Second
Mercury at Perihelion	56	35
Mercury at Aphelion	37	23

Considering the nearness of Mercury to the Sun, the Sun must be a comparatively large object in its sky. More than that, the size of the Sun and the quantity of light and heat Mercury receives must change enormously according to whether the planet is near the aphelion end of its orbit or the perihelion end (see Table 41 and Figure 31).

Table 41
THE SUN IN MERCURY'S SKY

	Sun		
	Apparent Diameter (minutes of arc)	Apparent Area (square minutes of arc)	Light Delivered (Earth = 1)
Earth	31.99	803.74	1.00
Venus	44.22	1,536.1	1.91
Mercury at Aphelion	68.3	3,653	4.54
Mercury at Perihelion	104.4	8,556	10.6

Even when Mercury is farthest from the Sun, it receives four and a half times as much heat and light as the Earth does, and over twice as much as Venus does. When Mercury is closest to the Sun, it receives over ten and a half times as much light as the Earth does and five and a half times as much as Venus does.

When the Sun was below the horizon, Mercury's night sky would be much like that which would be visible from the Moon, since there is no atmosphere on either world. The stars would not twinkle and all the objects in the sky would be about 30 percent brighter than in Earth's sky, because there would be no atmospheric absorption of light.

Of course, from the Moon we can see the Earth as a large globe in the sky, and there would be nothing like that in Mer-

Figure 31
THE SUN IN MERCURY'S SKY

SUN AS SEEN
FROM EARTH

SUN AS SEEN
FROM VENUS

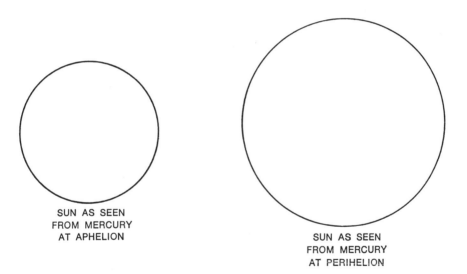

SUN AS SEEN
FROM MERCURY
AT APHELION

SUN AS SEEN
FROM MERCURY
AT PERIHELION

cury's sky. Since Mercury has no satellite, any more than Venus has, there is nothing in Mercury's sky like our Moon, either.

There is, however, Venus itself. Venus is not an evening star or morning star in Mercury's sky, but can be seen at any time during the night and at any distance from the Sun. What's more, the lighted side of Venus would be seen from Mercury when Venus was high in the sky.

When Mercury is at aphelion and Venus is closest in the sky, Venus is just over 1 minute of arc wide, about as large as we see it from Earth when it is at its nearest. From Earth, however, it is Venus's dark side that is facing us when it is nearest; from Mercury it is Venus's lighted side that would be visible. Venus would be seen with a magnitude of up to –6.7, or two and a half times as bright as we ever see Venus from Earth.

7

The Details of Mercury

MERCURY'S MASS

Mercury lacks satellites, so that it was hard in the past to determine its mass with any degree of accuracy. Now that we are able to send probes past it, however, the situation has improved. In Table 42 we have Mercury's mass as compared

Table 42
MASS OF MERCURY

	Mass	
	Moon = 1	Mercury = 1
Moon	1.00	0.22
Titan	1.90	0.42
Ganymede	2.03	0.45
Mercury	4.53	1.00
· Venus	66.3	14.6
Earth	81.3	17.9

with Venus and Earth, as well as with the large satellites that resemble it in diameter.

Mercury is over twice as massive as either Saturn's Titan or Jupiter's Ganymede, even though those satellites are greater in diameter than Mercury is. In fact, although it is frequently pointed out that there are two or three satellites that are larger than Mercury, the smallest planet, that "larger" refers only to size. Where mass is concerned, Mercury is considerably larger than any satellite, even the largest. Mercury is a small planet, just the same, as it is only 1/18 as massive as Earth.

The fact that Mercury is surprisingly massive as compared to Titan or Ganymede means that an unusual amount of mass is squeezed into each bit of volume of Mercury. Mercury, in other words, has a high density (see Table 43).

Table 43
DENSITY OF MERCURY

	Density	
	Kilograms per Cubic Meter	Pounds per Cubic Foot
Ganymede	1,900	120
Titan	2,000	125
Moon	3,400	212
Venus	5,200	325
Earth	5,518	345
Mercury	5,600	350

Mercury, Venus, and Earth are the only large bodies of the solar system to have densities of over 5,000 kilograms per cubic meter (310 pounds per cubic foot). All three, therefore, are bodies with a dense core that is probably chiefly or entirely metallic. And since Mercury is, by a small margin, the densest of the three, its core must be the largest in comparison to its total structure.

Mercury's great density gives it a high surface gravity compared to objects such as the large satellites, but this is not great enough to overcome its lack of total mass. Its surface gravity is low compared to Venus and Earth (see Table 44).

Table 44
SURFACE GRAVITY OF MERCURY

	Surface Gravity	
	Earth = 1	Mercury = 1
Titan	0.11	0.28
Ganymede	0.17	0.43
Moon	0.17	0.43
Mercury	0.39	1.00
Venus	0.88	2.26
Earth	1.00	2.56

Mercury's great density shows up if its surface gravity is compared with that of Mars, which is a larger planet with 2.7 times the volume of Mercury and 1.9 times the mass. Mars, however, is largely rock in structure, with only a comparatively small metallic core, so its density is only 3,950 kilograms per cubic meter (250 pounds per cubic foot), less than 3/4 that of Mercury. The result is that Mercury's surface gravity is almost exactly that of Mars, both being 2/5 that of Earth. A person weighing 70 kilograms (154 pounds) on Earth would weigh 27 kilograms (59 pounds) on either Mercury or Mars.

Mercury's density also shows up in its escape velocity (see Table 45). Here Mars's greater diameter plays a role in mak-

Table 45
ESCAPE VELOCITY OF MERCURY

	Escape Velocity	
	Kilometers per Second	Miles per Second
Titan	2.3	1.4
Moon	2.38	1.48
Ganymede	2.9	1.8
Mercury	4.3	2.7
Venus	10.3	6.4
Earth	11.2	7.0

ing its escape velocity somewhat higher than that of Mercury. Mars's escape velocity is 5.0 kilometers per second (3.1 miles per second), or about 1 1/6 that of Mercury.

MERCURY'S ROTATION: OLD STYLE

What about the rotation period of Mercury? Would there be a problem in discovering it because of an obscuring cloud layer as in the case of Venus?

The smaller the surface gravity of a world, the less able that world is to hold an atmosphere. The Moon lacks sufficient gravitational pull to hold an atmosphere, for instance. Mars, which has a surface gravity 2.5 times that of the Moon, manages to hold a thin atmosphere about 1/100 as dense as that of Earth. In that case, since Mercury's surface gravity is almost exactly equal to that of Mars, ought we to expect that it, too, could hold a thin atmosphere?

The answer is, No.

The ability of a planet with a given gravitational field to hold an atmosphere depends also upon its temperature. As the temperature goes up, the ability to hold an atmosphere goes down. Mercury is, on the average, only 1/4 the distance from the Sun that Mars is, so that Mercury receives, on the average, 16 times the light and heat that Mars does. Mercury's surface temperature would be expected to be much higher than that of Mars, therefore, and Mercury's ability to hold an atmosphere would be much less.

All this was well understood by the 1870's, and astronomers were therefore quite sure that Mercury not only lacked a cloud layer, but that it lacked any significant atmosphere as well.

In that case, should we not be able to determine Mercury's rotational period? It would only be necessary to note a feature on Mercury's bare surface and follow it around the planet.

Easier said than done! Mercury is so small a planet and so distant and (worse yet) always so near the Sun that detailed observation of its surface by telescope from Earth is very difficult indeed.

To be sure, some astronomers thought they could make out mountains on Mercury, and in the early 1800's there were suggestions, based on such observations, that Mercury (like Earth and Mars) rotated in a period of about one day.

Such suggestions were not taken seriously by most astronomers. For one thing, Mercury's shape seemed quite spherical —there was no perceptible oblateness. This meant it could not be rotating quickly. Then, too, Mercury is so near the Sun that astronomers felt reasonably certain that the Sun's tidal influence would force Mercury into a gravitational lock, so that it would turn one face always to the Sun, just as Earth's tidal influence forces the Moon to turn one face always to Earth. The reasonable guess, then, was that Mercury's rotational period was equal to its period of revolution—that is, 88 days. What's more, the rotation would have to be counterclockwise or "direct," if the gravitational lock was to be maintained.

With this in mind, the Italian astronomer Giovanni Virginio Schiaparelli studied Mercury carefully, and finally made out vague patterns of dark and light on the lighted portion of Mercury when it was at maximum elongation in the evening. If its period of rotation were 88 days, that visible pattern should remain the same at each maximum elongation. It seemed to do so, and in 1889 Schiaparelli announced that Mercury's period of rotation was indeed 88 days.

Other observers confirmed Schiaparelli's findings in hundreds of careful observations, and for 75 years there was no doubt as to the matter.

If Mercury were gravitationally locked in this fashion, with one side always facing the Sun, that Sun side would lie under a perpetual swollen Sun that would expand for 44 days as Mercury traveled from aphelion to perihelion, until that Sun was nearly 11 times as bright as when seen from Earth. The Sun would then shrink for 44 days down to a mere 4 times Earth brightness as Mercury traveled from perihelion back to aphelion. We might expect that Mercury's Sun side, under such conditions, would blaze into a temperature that would be maintained at a level well beyond that required to melt lead.

As for Mercury's night side, it would never see the Sun and it would cool down to the average temperature of space itself, about –270° Celsius (–454° Fahrenheit), or only three Celsius degrees above absolute zero. Mercury's night side would be colder than the surface of any other planet, even distant Pluto, for all that it lay so near the Sun.

This picture of Mercury was so dramatic that science-fiction writers used it frequently. In fact, there was an assumption sometimes that vast piles of frozen air might exist on the night side, and that it might be mined by visitors from Earth.

This, however, cannot be so. If Mercury were now locked into position in its orbit about the Sun, with one side always facing the Sun, there must have been a period of time before the lock was established, when Mercury rotated more quickly and when all parts of it received sunlight at one time or another. Mercury's rotation would have slowed rapidly (as astronomical time goes), but there would have been plenty of time for all its atmosphere (what there was of it to begin with—and there might have been none even at the start) to vanish into space. We would expect, therefore, that the night side would be as airless as the Sun side.

Another common science-fictional view of Mercury was that there was a "twilight zone." Between the Sun side and the night side, might there not be a boundary area within which the temperature would be moderate and human habitation be possible without too much difficulty?

Actually this, too, cannot be so. For one thing, it would work best if there were an atmosphere and an ocean to absorb sunlight and to act as a heat reservoir. Human beings could, in that case, retreat just within the night-side boundary and allow the warmed air and water coming from the Sun side to make the temperature of their surroundings mild. Without air and water there would be a sharp boundary, and even a bit of Sun at the horizon might well be uncomfortably warm, while even if the Sun were just below the horizon it would be uncomfortably cold.

Worse yet, there would be no sharp, motionless boundary between the Sun side and the night side. This can be explained as follows.

If Mercury rotated in 88 days, it would do so at a perfectly even rate. There would be one complete turn (360 degrees) in 88 days or 2,112 hours. That means that each hour the Sun would seem to drift from east to west in Mercury's sky for just 1/2,112 of 360 degrees, or 10.23 minutes of arc.

If Mercury orbited the Sun in a perfect circle, then it would move at a constant speed, counterclockwise, about the Sun, and the Sun would drift from west to east in Mercury's sky at a rate of 10.23 minutes of arc per hour.

The two drifts, rotational and revolutionary, would exactly cancel each other so that Mercury's Sun would appear to be completely stationary in the sky. In that case, there would be a sharp and permanent boundary between Sun side and night side.

Mercury, however, does *not* move in a perfect circle. It moves in an ellipse, and a fairly eccentric one. As Mercury approaches perihelion, it moves faster and faster, moving fastest at perihelion. It then slows as it moves toward aphelion, moving slowest at aphelion. It then speeds up and goes through the cycle over and over.

In the perihelion half of its orbit, the west-to-east drift of the Sun in Mercury's sky due to its revolution about the Sun is faster than the east-to-west drift of the perfectly steady rotation. In the aphelion half of its orbit, the situation is reversed.

This means that during the perihelion half of the orbit, the Sun drifts slowly eastward in the sky, and during the aphelion half of the orbit slowly westward again, over and over.

The back-and-forth motion of the Sun in Mercury's sky is completed in just 88 days. The motion eastward takes somewhat less than 44 days, and the return motion westward takes somewhat more than 44 days, because Mercury spends more time in the slow-moving aphelion half of the orbit than in the fast-moving perihelion half.

This back-and-forth motion is like that of a pendulum or the up-and-down motion of a scale used for weighing objects. It is called "libration" from the Latin word for "scale."

Because the Sun moves in Mercury's sky like this, there can be no sharp boundary between Sun side and night side. If you were where you thought the sharp boundary might be, the Sun would rise and set in an 88-day cycle, moving quite a way above the horizon and moving quite a way below it.

The surface of Mercury would therefore be divisible into four parts. There would be a true Sun side, taking up less than half the total surface, in which the Sun, while wobbling, would remain always in the sky. There would be a night side directly opposite, in which the Sun would never appear.

Between the two there would be two "libration zones," one on each side of the planet, within which the Sun would be in the sky part of the time and absent part of the time. These libration zones would not be mild-tempered "twilight zones." They would undergo extreme temperature changes and no part of Mercury would be comfortable.

MERCURY'S ROTATION: NEW STYLE

The first hint that there might be something wrong with the notion of Mercury's rotating in 88 days came in 1962. In that year the night side of Mercury was found to be emitting micro-waves of a wavelength and intensity that seemed to indicate that the temperature of the surface was quite high.

This simply couldn't be so if Mercury rotated in 88 days, since in that case the center of the night side would never see the Sun and would be frigid indeed.

The matter could be settled by bouncing a beam of microwaves from Mercury's surface and determining the speed of rotation of Mercury from the changes observed in the reflected beam.

The results were a complete surprise. In 1965 two electrical engineers, Rolf Buchanan Dyce and Gordon H. Pettengill, working with microwave beam reflection, announced that Mercury did *not* rotate in 88 days, but in about 59 days. The result was checked out by others and also by probes that eventually reached the neighborhood of Mercury. We now know that the rotation period of Mercury relative to the stars (Mercury's "sidereal day") is 58.65 days (see Table 46).

There are thus three slowly rotating bodies in the inner solar system and two quickly rotating bodies, something that can be shown most dramatically if we consider the equatorial speeds of rotation relative to the stars (see Table 47).

Table 46
SIDEREAL DAY OF MERCURY

	Length of Sidereal Day		Direction of Rotation	Axial Inclination (degrees)
	Earth Days	Hours		
Mercury	58.65	1,407	Direct	7.0
Venus	243.09	5,834	Retrograde	183.4
Earth	0.997	23.93	Direct	23.45
Moon	27.322	656	Direct	6.68
Mars	1.026	24.62	Direct	23.98

Table 47
EQUATORIAL SPEED OF MERCURY

	Equatorial Speed of Sidereal Rotation	
	Kilometers per Hour	Miles per Hour
Mercury	10.82	6.72
Venus	6.54	4.06
Earth	1,674.4	1,040.5
Moon	16.65	10.35
Mars	866.4	538.4

A sidereal period of 58.65 days is just 2/3 of 88 days, so that the sidereal day of Mercury is 2/3 of the Mercurian year. It turned out, when astronomers did their calculations, that the Sun's tidal effect could indeed freeze Mercury into this situation. That could have been calculated before the discovery of Mercury's rotation rate, but there seemed no reason to look in that direction. The full-year day is a more efficient way of freezing Mercury's rotation, and that had seemed to be the answer.

How is it, though, that Schiaparelli and all the others who backed his observation came to make their mistake? Why did they think the rotational period was 88 days?

Since Mercury rotates in 58.65 days, that means that the east-to-west drift of the Sun due to the planet's rotation is 15.345 minutes of arc each hour (one and a half times what it would have been if the rotation had been slower and had taken 88 days). The revolution of Mercury about the Sun produces a west-to-east drift of an average of 10.23 minutes of arc per hour.

If we combine the two, then the Sun drifts from east to

west at a rate of 5.115 minutes of arc per hour on the average. This means that the Sun doesn't stay in one spot in the sky, nor does it do no more than oscillate back and forth in a libration. The Sun moves from east to west in Mercury's sky more or less steadily. Wherever you are on Mercury (except perhaps very near the poles), the Sun rises in the east and eventually sets in the west, and then later on rises in the east again, over and over—just as on Earth. There is no Sun side and no night side. Every part of Mercury experiences day and night.

The Sun moves more slowly in Mercury's sky than in Earth's, of course. An east-to-west drift of 5.115 minutes of arc per hour means that the Sun will make a complete circle of the sky in 4,222.9 hours, or 176 Earth days. This is the time from sunrise to sunrise on Mercury and the length of its "solar day"; it is the longest solar day among the planets (see Table 48).

Table 48
SOLAR DAY OF MERCURY

	Length of Solar Day	
	Earth Days	Hours
Mercury	176	4,224
Venus	116.78	2,802.65
Earth	1.00	24.00
Moon	29.53	708.7
Mars	1.03	24.66

Notice that 176 days is just twice the 88-day period of Mercury's revolution. Suppose, then, that Mercury is at perihelion. One side (call it A) faces the Sun at that time. In 88 days, Mercury revolves about the Sun and is back at perihelion. In

88 days, however, Mercury has only completed half a rotation with respect to the Sun. Side A now faces *away* from the Sun. When Mercury completes a second revolution, however, Side A faces toward the Sun again. The same argument holds for every other position of Mercury's orbit. In each position, the same side faces the Sun once every other revolution.

This holds for Mercury at maximum elongation. Suppose we study Mercury at every maximum elongation west of the Sun (in the dawn). We would see the same side of Mercury not every time, but every other time. There are occasions, though, when every other maximum elongation is easy to observe while the ones in between are not, because long and short maximum elongations (depending on whether Mercury is near aphelion or near perihelion) alternate. Astronomers studying every other elongation see the same side, the same pattern of light and dark, each time. It is easy to assume, then, that Mercury *always* shows the same side to the Sun and therefore that it rotates relative to the stars in 88 days.

It is an easy mistake to make, and if it weren't for microwave reflections and rocket probes, astronomers might never have learned better. Now that they have, however, careful observation of Mercury from Earth shows that the same side is not always presented to the Sun at every point in the orbit.

If Mercury moved about the Sun in a perfect circle, the Sun would seem to move across its sky at an even east-to-west rate of 5.115 minutes of arc per hour. There is, however, Mercury's orbital eccentricity, which makes its speed of revolution vary from point to point in the orbit and causes the Sun's apparent motion to speed up and slow down. In fact, when Mercury is moving fastest, near perihelion, the slowdown is so great that for a time, the Sun actually slips into a brief retrograde motion from west to east.

This produces some remarkable results.

Suppose an observer is on a spot on Mercury's surface that happens to have the Sun directly overhead when it is at perihelion.

The Sun will rise in the east, while it is actually farthest from Mercury. It is then a little more than twice the width of the Sun as seen from Earth, and four and a half times as hot. As it rises (which it does slowly, for it is well over two months from sunrise to sunset), it grows larger. By the time it is at zenith, it is well over three times the width of the Sun as seen from Earth and nearly eleven times as hot.

Nor does the Sun pass the zenith quickly. Its westward motion has slowed steadily as it approaches the zenith. A small way past the zenith it comes to a momentary halt and begins to move *eastward* very slowly. After passing the zenith again, it comes to another momentary halt, then begins to move westward again. This time it moves steadily, speeding up and shrinking in size as it moves toward the western horizon. When it sets, it is at aphelion again, and at minimum size. On a day such as this, there are no fewer than three noons.

On the directly opposite side of Mercury, the same thing happens. During one of Mercury's revolutions, one side gets it; during the next, the other side does; and so on in alternation.

In between those two opposing points over which the perihelion Sun passes at zenith, there are points where the Sun is at maximum size somewhere between zenith and the horizon. Then it does its dance, backward and forward, at those points.

The most dramatic places on Mercury, perhaps, are those where the Sun is at perihelion at the horizon.

At such a point a swollen Sun will rise in the east, move above the horizon more and more slowly, pause and, as though it can't face the world, begin to sink and then set. After a while it rises a second time, and this time it doesn't change its mind.

It moves toward the zenith more and more quickly, shrinking as it goes. At zenith, Mercury is at its aphelion and the Sun is down to minimum size.

The Sun sweeps past zenith without turning and begins to grow again as it sinks toward the west. It sinks more and more slowly and swells as it does so. By the time it approaches the horizon, it is expanded to nearly full size. It sets and, after a pause, rises again as though to take a last look around to make sure all is well. Then it sets and is gone for many days.

The temperature of Mercury, when the Sun is in the sky, varies somewhat according to the position of the Sun and exactly where it is at its most swollen. The maximum temperature is about 425° Celsius (800° Fahrenheit). (See Table 49.)

Table 49
TEMPERATURE OF MERCURY

	Maximum Temperature		Minimum Temperature		Temperature Range	
	°C	°F	°C	°F	°C	°F
Mercury	425	800	−180	−290	605	1090
Venus	475	890	475	890	−	−
Earth	57	135	−52	−125	109	260
Moon	102	215	−175	−280	277	495
Mars	30	85	−150	−240	180	325

This is not as hot as it is on Venus, even though Venus is nearly twice as far from the Sun and has a cloud layer that reflects three fifths of the sunlight that falls upon it. However, Venus has an atmosphere that traps and holds the Sun's heat, and Mercury doesn't.

Then, too, Venus's atmosphere not only traps the heat but spreads it over the entire planet, so that Venus remains extremely hot always, whether by day or by night.

Mercury, on the other hand, has no atmosphere and cools down rapidly after the Sun sets. By dawn, after having experienced the absence of sunlight for well over two months, Mercury's surface has cooled to a low point of −180° Celsius (−290° Fahrenheit). The extreme difference in temperature on Mercury between day and night is about 605 Celsius degrees, or 1090 Fahrenheit degrees.

MERCURY'S SURFACE

Although microwave reflection told astronomers the surprising story of Mercury's rotation, it still couldn't give them sufficient detail about the nature of the surface. Since no cloud layer hid that surface, a rocket probe would be ideal for the purpose.

On November 3, 1973, Mariner X was launched. It passed by the Moon and then, on February 5, 1974, it passed by Venus just 5,800 kilometers (3,600 miles) above the cloud layer and sent back useful data on that planet. It then headed for Mercury, and on March 29, 1974, passed within 700 kilometers (435 miles) of the surface.

It moved into an orbit about the Sun in such a way as to make one circuit in 176 days, or just twice Mercury's year. This brought it back to Mercury in the same spot as before, because each time the probe made one circuit of the Sun, Mercury was making two. Mariner X passed Mercury on September 21, 1974, a second time, and then on March 16, 1975, a third time. On the third pass, it skimmed within 327 kilometers (203 miles) of Mercury's surface.

After the third pass, Mariner X had consumed the gas that

kept it in a stable position, and it was thereafter useless for further study of the planet.

Mariner X confirmed Mercury's rotation rate and temperature, and showed that it had no satellite and no significant atmosphere. It determined its diameter, mass, and density.

The photographs it took of Mercury showed a landscape that looked very much like that of the Moon. There were craters, craters, everywhere, with the largest one about 200 kilometers (125 miles) in diameter.

On the whole, Mercury had fewer craters than the Moon per unit area, particularly larger craters. This may be because Mercury's stronger gravitational field prevented meteor collisions from making such large splashes.

The Moon, especially the side that faces us, has large *maria,* or "seas." These are relatively flat basins that early in the Moon's history must have formed as lava flows. Mercury is not as rich in basins as the Moon is. The largest one sighted is about 1400 kilometers (870 miles) across. It is called "Caloris" ("Heat"), because it is just about at the spot on Mercury that is under the Sun at zenith when Mercury is at perihelion.

Mercury also possesses long scarps, or cliffs, that are several hundred kilometers long and about 2.5 kilometers (1.5 miles) high.

Mariner X only photographed about 3/8 of the surface of Mercury in the course of its three working passes. Astronomers imagine that the remainder of the planet's surface is much like that which we've seen, but we can never be sure. The solar system has been too full of surprises in the last thirty years for astronomers to take anything for granted.

For instance, Mariner X on its first pass discovered that Mercury had a small magnetic field. It was only a hundredth as intense as Earth's magnetic field, but it was there.

For an astronomical object to have a magnetic field, there are two requirements according to current theory. There must be a planetary core of some liquid capable of conducting an electric current, and there must be a rotation fast enough to set up swirls in the internal liquid.

Mercury almost certainly has a liquid core of the right type, but it does not rotate nearly quickly enough to set up a magnetic field—yet it has one. Astronomers haven't quite figured that out yet.

8

Asteroids

Is there any object other than Mercury with an orbit about the Sun that lies entirely within that of Venus?

During the last half of the nineteenth century, astronomers thought there was indeed such an object, one that was not only nearer to the Sun than Venus is, but even nearer than Mercury is. The story starts with Mercury.

If we imagine a line crossing the orbit of a planet, from its perihelion through the Sun to the aphelion, that is the major axis of the ellipse. If the planet and the Sun were all there was to consider, then that major axis of the ellipse would be fixed in space. The planet would repeat its path exactly, over and over again.

The planet is, however, affected to a small degree by the weak gravitational pulls of other planets. Mercury, for instance, feels the pull of Venus, of Earth, and even of distant Jupiter. The effect of these pulls is to cause the major axis to twist in space very slowly. Each time Mercury comes to its perihelion, that perihelion has moved a very small amount in the counterclockwise direction from where it was before, by an amount equal to 1.38 seconds of arc (see Figure 32).

In actual distance, it means that Mercury must travel about 280 additional kilometers (175 additional miles) before catching up to the perihelion—a distance it covers in five seconds. At this rate, the perihelion advances about 574 seconds of arc per century and would make a complete circle of Mercury's orbit in 225,784 years.

The trouble is that when astronomers took into account the pull of the various planets on Mercury, they could only account

Figure 32
ADVANCE OF MERCURY'S PERIHELION

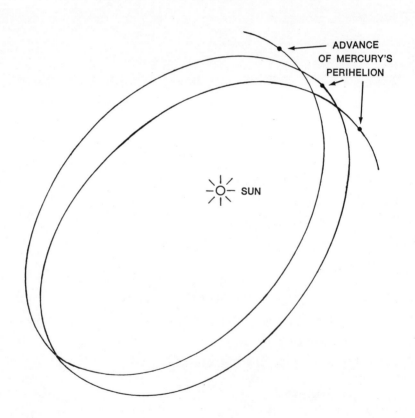

for an advance of 531 seconds of arc per century. Mercury's perihelion turned out to be advancing 43 seconds of arc per century faster than astronomers could account for.

That is not much, but it shouldn't be there, and astronomers were very puzzled. There were three possibilities that they could see. The first was that something was wrong with the theory of gravitation. That just didn't seem as though it could be. The theory of gravitation had been checked any number of times since it was first advanced, and no error had ever been found.

The second possibility was that the gravitational influence of one or more of the other planets had been wrongly calculated. If Venus, for instance, had 10 percent more mass than it was thought to have, it would have a slightly stronger pull on Mercury and that would increase the rate of advance of Mercury's perihelion till it was just right.

However, if Venus were 10 percent more massive than was thought, it would also have a stronger pull on Earth and that would affect Earth's motions very slightly—and no such very slight effect could be observed. The same would be true if the mass of any other planet were adjusted: the effect would be not only on Mercury but on other planets. And, indeed, astronomers are now quite sure of the masses of the known planets and these do *not* account for that extra 43 seconds of arc per century.

That left a third possibility.

What if there were a gravitational pull that was not being taken into account at all? What if there were an undiscovered planet close enough to Mercury and large enough to affect it?

If there were a planet between Venus and Mercury, even if it was quite small it would surely have been seen by astronomers. Then, too, if it were large enough to affect Mercury it would also be large enough to affect Venus, and there was nothing in Venus's motions that couldn't be explained by the objects that were already known.

But what if there were a small planet even closer to the Sun than Mercury was? It might be large enough and close enough to Mercury to affect it, but too distant from Venus (and certainly from the planets beyond Venus) to produce much of an effect there. Then, too, it would always be so near the Sun that it would be hard to detect at all from Earth.

Beginning in 1855, astronomers began to consider the problem of the advance of Mercury's perihelion. The French astronomer Urbain J. J. Leverrier (luh-vehr-YAY) felt there were two ways in which such a planet, very close to the Sun, might be detected.

First, when there was a total eclipse of the Sun this innermost planet might be detected. So close to the Sun, it would always be drowned in the Sun's light except during an eclipse. Second, it might occasionally pass in front of the Sun in a transit, and then it would be visible as a tiny black dot moving across the Sun's face. The only three bodies that ever did that were Mercury, Venus, and our Moon, and if none of them was scheduled for a transit, the dot would have to represent a new planet.

Leverrier began to search astronomical records to see if anyone had reported something that looked like a planetary transit across the Sun where none had been scheduled. In 1859 a French amateur astronomer named Lescarbault (luh-scar-BOH) reported having observed such an event. Leverrier interviewed him, checked other reports, and finally decided that there was indeed a planet closer to the Sun than Mercury.

Using Lescarbault's data and other data he had found in the records, Leverrier announced the new planet, which he named "Vulcan." This was a suitable name, for Vulcan was the Roman god of the forge, working always at the hot fire to produce metallic masterpieces. The Sun is certainly the forge of the solar system, and the planet Vulcan was always in its neighborhood.

Leverrier, from the data he could find, judged Vulcan to have a diameter of about 1900 kilometers (1200 miles). This would give Vulcan a diameter only 5/9 that of the Moon, a surface area of 11,700,000 square kilometers (4,200,000 square miles), or just a little more than that of Europe, and a volume about 1/6 that of the Moon (see Figure 33).

If one supposed it to have a density like that of Mercury, its mass would be about 3/10 that of the Moon.

Leverrier further calculated that its average distance from the Sun would be 21,000,000 kilometers (13,000,000 miles), which is only about 1/3 the average distance of Mercury from the Sun (see Figure 34). If this were so, it would never be seen more than 8 degrees from the Sun. It would never set in the evening more than half an hour after the Sun, or rise in the morning more than half an hour before the Sun.

What's more, assuming it was without an atmosphere and

Figure 33
"VULCAN"

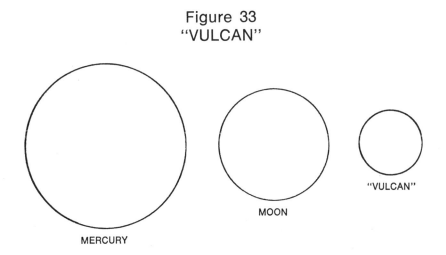

MERCURY

MOON

"VULCAN"

Figure 34
ORBIT OF "VULCAN"

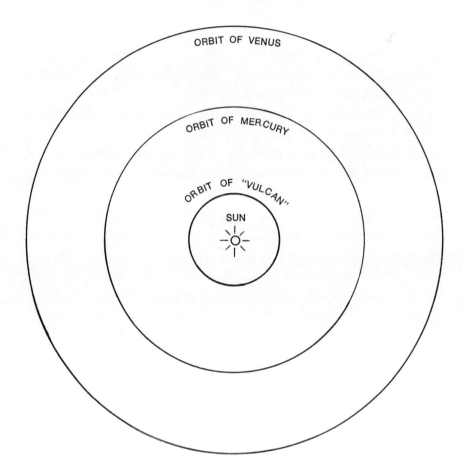

had a surface like that of Mercury or our Moon, it would have a maximum magnitude of +0.5. It would be only 1/5 as bright as Mercury at its brightest, and it would be seen under much worse conditions. The chance of its being seen at all would be virtually zero, except during eclipses or when it was in transit.

Leverrier estimated its orbital period to be 19.7 days, only a little over a fifth that of Mercury.

At Vulcan's distance from the Sun, the Sun's tidal influence on Vulcan would be about 15 times that on Mercury, allowing for its greater proximity to the Sun (which would increase the tidal influence) and its smaller size (which would decrease the tidal influence). It would be fairly certain, given that condition, that Vulcan would face one side only to the Sun at all times. It would have the Sun side and the night side that Mercury was thought to have for so long and proved not to have.

From Vulcan's distance, the Sun would appear to have a diameter of 3.8 degrees, about 7 1/2 times its diameter as seen from the Earth. The Sun would deliver 55 times as much light and heat to Vulcan as it delivers to the Earth, and 5 times as much as it delivers to Mercury at perihelion. It seems quite likely that the Sun side of Vulcan would be glowing with a dim red heat.

What Leverrier had done was only theory, of course. Now that his calculations had been made, it would be necessary actually to detect the planet. Leverrier used his calculated orbit to predict when Vulcan would be in transit across the Sun, but such transits were never seen. Instead, other events that might be transits were reported.

Leverrier died in 1877 still convinced that Vulcan existed. In 1878, when a solar eclipse was visible in the western United States, many astronomers were in its path with their telescopes in order to try to spot Vulcan. Some reports of possible sightings were made, but they weren't convincing.

By now, though, the art of photography was being applied to astronomy, which no longer had to depend on the eye alone. Careful photographs were taken of the vicinity of the Sun on the occasion of every total eclipse, and these could be studied in detail and at leisure.

In 1900 the American astronomer Edward Charles Pickering announced that there couldn't possibly be any object circling the Sun within Mercury's orbit that was brighter than +4 in magnitude. In 1909 another American astronomer, William Wallace Campbell, maintained that there could be nothing inside Mercury's orbit with a magnitude higher than +8.

If that were so, it would mean there was nothing with an orbit smaller than that of Mercury that could be more than 48 kilometers (30 miles) in diameter. It would take a million bodies of that size to account for the movement of Mercury's perihelion.

Everything we have learned since then, including the use of planetary probes, confirms that. Leverrier was wrong. There is *no* planet Vulcan and Figures 33 and 34, which I have presented of Vulcan and its orbit, represent nothing more than an imaginary body that doesn't exist. There is absolutely nothing that we have observed up to this very moment that moves around the Sun in an orbit entirely within that of Mercury.

But that 43-seconds-of-arc advance of Mercury's perihelion that couldn't be explained in Leverrier's time is still there. If there is no planet Vulcan, and nothing at all inside Mercury's orbit, and if the masses of the known planets are what we now know they are, then there is only one alternative left: something is wrong with Newton's theory of gravity.

And, indeed, that turns out to be the case.

In 1916 the German-born scientist Albert Einstein advanced his "general theory of relativity," which produced a new version of gravity. It worked out to be almost the same as Newton's, but not quite. In Einstein's theory, for instance, energy is shown to be the equivalent of mass, though it takes a huge amount of energy to be equivalent to a small amount of mass.

The Sun's gravitational field is a form of energy, and this is equivalent to a certain amount of mass, which produces a bit of gravitation of its own. In other words, the Sun's gravitational field has a (much smaller) gravitational field which must be added to that of the Sun. This additional bit of gravitation, if taken into account, just explains that extra advance of Mercury's perihelion. There is no need to drag in Vulcan.

Of course that extra bit of gravitation also affects Venus and all the other planets, but less and less as distance increases. Only in the case of Mercury is the effect really noticeable.

EARTH-GRAZERS

Since there is nothing circling the Sun with an orbit completely inside that of Venus, does that mean there is nothing left to talk about in this book?

Not quite. There are small objects with elongated orbits. These orbits do not lie within that of Venus altogether, but their perihelia do. In other words, there are objects other than Mercury which through *part* of their orbits are closer to the Sun than Venus is.

It was not until the beginning of the 1800's that the existence of small objects circling the Sun became known. Before then, the small objects that had been sighted were always satellites circling one planet or another.

On January 1, 1801, however, a new planet was discovered by the Italian astronomer Giuseppe Piazzi (PYAH-tsee). It had an orbit that lay between those of Mars and Jupiter. Piazzi named it Ceres.

The reason it had not been discovered sooner was that it was much smaller than any of the other known planets, and smaller even than most of the satellites then known. Its diameter is about 1,000 kilometers (625 miles)—only about 1/5 that of Mercury and 1/13 that of Earth. The volume of Ceres is about 1/112 that of Mercury and about 1/2000 that of Earth.

But then came something more startling. Within six years,

three more planets were discovered with orbits between those of Mars and Jupiter. They were named Pallas, Juno, and Vesta and each one was even smaller than Ceres. The smallest was Juno, with a diameter of about 240 kilometers (150 miles).

Not only were these planets much smaller than the other planets, but their orbits were not as carefully arranged. Their orbits, when drawn on paper, seemed to cross. (They didn't really cross, because their orbits moved in three dimensions and each one, where they seemed to cross, actually passed well above or well below the orbit of another.)

The orbital inclinations were all higher than that of any of the known planets, ranging from 7.1 degrees for Vesta to 34.8 degrees for Pallas. Their orbital eccentricities were high, too. For both Pallas and Juno they were higher than that of Mercury.

It seemed that some special name had to be applied to these four very unusual bodies. The German-English astronomer William Herschel pointed out that they were so small that through the telescope they did not expand to form visible orbs as the other, larger planets did. Instead they remained simple points of light, like stars. He suggested that they be called "asteroids" (from a Greek word meaning "starlike"), and the name has stuck.

Those four weren't the only asteroids to be discovered. As the years passed, more and more were discovered, especially after astronomers began to use photography for the purpose.

By 1898, after nearly a century of discoveries, no fewer than 432 asteroids had been discovered. Of these, every one had an orbit that lay between those of Mars and Jupiter. The region between the orbits of those two planets became known as the asteroid belt.

Might there be asteroids outside the asteroid belt? Astronomers would not be surprised if there were asteroids beyond

Jupiter, since small bodies would be very difficult to detect at such great distances. It did not seem likely that asteroids would venture out of the belt in the other direction and have orbits within that of Mars. They would then be sufficiently close to Earth to be easily detected and as 1898 opened, no such asteroids had been found.

Then, on August 13, 1898, a German astronomer, Gustav Witt (Viht), discovered the 433rd asteroid. When its orbit was calculated there was cause for astonishment, for its period of revolution was the first one discovered for an asteroid that was less than that of Mars. The period of revolution of Asteroid #433 was only 1.76 years, or 643 days, which is 44 days less than that of Mars, and is much less than the 2.7-year period that is average for an asteroid.

What's more, Asteroid #433 had an orbital eccentricity of 0.223, which was not unusually high for an asteroid but which, combined with its small orbit, meant that its perihelion must be well within the orbit of Mars.

It turned out that at perihelion, Asteroid #433 approached within 170,000,000 kilometers (105,000,000 miles) of the Sun, an approach not very much greater than that of Earth. In fact, at perihelion it is closer to Earth's orbit than to Mars's orbit, and it was therefore the first asteroid ever discovered whose orbit carried it outside the asteroid belt. It wasn't outside the belt all through its orbit, however, for its aphelion was well within the belt.

It turned out that if Asteroid #433 and Earth were each at the proper point in their orbits, the distance between them could be as little as 23,000,000 kilometers (14,000,000 miles). This is only a little over half the minimum distance of Venus from Earth and it means that, if we don't count our own Moon, Asteroid #433 becomes our nearest neighbor in space.

Witt named this new and unusual asteroid after the Greek

god of love, Eros. It was the first time an asteroid had been given a masculine name, and from that time all asteroids that venture outside the asteroid belt have been given such masculine names. All others receive feminine names.

Any asteroid that can approach Earth more closely than Venus does came to be called an "Earth-grazer." Eros was the first of the Earth-grazers to be discovered.

In 1931 Eros approached to a point only 26,000,000 kilometers (16,000,000 miles) from Earth. In that position it had an unusually large parallax, and its pointlike appearance made it easy to determine that parallax accurately. A vast astronomical effort to determine the parallax was launched under the English astronomer Harold Spencer-Jones. As a result, the distances of the solar system were determined more accurately than ever before. It was only when microwave beams came into use that the Eros observations were improved on.

Between 1898 and 1932, only three more Earth-grazers were discovered, and each of those three approached Earth less closely than Eros did.

The record was broken, however, on March 12, 1932, when a Belgian astronomer, Eugene Delporte, discovered Asteroid #1221 and found that its orbit gave it a perihelion distance of 162,000,000 kilometers (100,000,000 miles) from the Sun, a little closer than that of Eros. Delporte gave the new asteroid the name "Amor," the Roman equivalent of Eros.

When Earth and Amor were each at the appropriate part of their orbits, they would be separated by only 16,000,000 kilometers (10,000,000 miles).

APOLLO-OBJECTS

On April 24, 1932, the German astronomer Karl Reinmuth discovered an asteroid he named Apollo. It was appropriate to name it after the Greek god of the Sun, for when the orbit was worked out this sixth Earth-grazer did not merely penetrate within Mars's orbit as the other five do. It went past the orbit of Earth, too, and even that of Venus, ending up on July 7, 1932, at a perihelion point only about 97,000,000 kilometers (about 60,000,000 miles) from the Sun. This is roughly 10,000,000 kilometers (6,200,000 miles) closer to the Sun than Venus gets.

Apollo's orbital eccentricity is 0.566, so its aphelion is 345,000,000 kilometers (214,000,000 miles) from the Sun. This means that at the far end of its orbit it is within the asteroid belt. Its elongation is such that although it approaches the Sun much more closely than Eros does, Apollo spends so much time in the asteroid belt that its period of revolution is longer than that of Eros. It is 1.81 years or 661 days, which is 18 days longer than that of Eros.

On May 15, 1932, Apollo was only 11,000,000 kilometers (6,800,000 miles) from Earth—a new record approach.

Apollo was lost shortly after it was observed, because in the time it was studied, the orbit worked out for it wasn't sufficiently accurate, and an object so small was bound to be so dim that if astronomers didn't know *exactly* where to look for it, it would be almost impossible to spot. In 1973, however, the American astronomer Brian Marsden and a group of colleagues made a concerted search in various possible places and located it.

Apollo was the first asteroid discovered that passes within Earth's orbit at perihelion, but it was not the last. All such objects are now called "Apollo-objects."

In February 1936, Delporte, who had detected Amor four years earlier, detected another object which he named Adonis. It seemed unusually bright for a tiny asteroid and moved unusually quickly, because it happened to be quite close to the Earth when it was discovered. (That is no coincidence, for the closer to the Earth a small asteroid is, the brighter it is, the more rapidly it moves, and the easier it is to discover it.)

Just a few days before it was detected, Adonis had passed only 2,400,000 kilometers (1,500,000 miles) from Earth, and had been just a little over 6.3 times the distance of the Moon from us.

What's more, the astonishing record set by Apollo for perihelion distance was broken, for at its closest approach Adonis is only 66,000,000 kilometers (41,000,000 miles) from the Sun, only two thirds the distance of either Apollo or Venus.

Indeed, with Adonis we can begin talking about the planet Mercury, for Adonis has a perihelion distance from the Sun that is slightly smaller than the distance of Mercury from the Sun at its aphelion. There are times, in other words, when Adonis can be up to 4,500,000 kilometers (2,800,000 miles) closer to the Sun than Mercury is.

As was true of Apollo, Adonis was not seen again after its discovery. Unlike Apollo, it has not yet been detected again.

Adonis's perihelion distance remained a record for a number of years, but its close approach to Earth did not. In November 1937, Reinmuth (the discoverer of Apollo) detected a third Apollo-object and named it Hermes. Hermes moved across the sky with extraordinary quickness and was gone almost before a stab could be made at calculating its orbit. At its perihelion it was 87,000,000 kilometers (54,000,000 miles) from the Sun, quite a bit closer to the Sun than Apollo ever reached.

It turned out, though, that Hermes had passed within 800,000 kilometers (500,000 miles) of Earth, and, if its orbit had been calculated correctly, it was possible for it to miss us by a mere 310,000 kilometers (190,000 miles), at which time its distance from us would be only 4/5 that of our Moon.

Hermes set a close-approach record for objects large enough to be visible outside our atmosphere, and no one is eager to have that record broken. Hermes, too, was lost after its fly-by and has never been seen again.

ICARUS

On June 26, 1949, an Apollo-object was accidentally discovered by the German-American astronomer Walter Baade, and this turned out to be the most unusual asteroid of all. For one thing, its period of revolution turned out to be only 1.12 years, or 409 days, the smallest for any asteroid up to that time. For another, its orbital eccentricity was 0.827, the largest up to that time for any asteroid.

A small period of revolution means a small orbit, and a large eccentricity involving such a small orbit must mean that the perihelion is quite close to the Sun. And so it is. At perihelion, the new asteroid is only 28,500,000 kilometers (17,700,000 miles) from the Sun. This is less than half the distance of Adonis, the previous record holder, and is only about 60 percent of the closest approach Mercury ever makes to the Sun (see Figure 35).

Naturally, Baade named the asteroid Icarus, after a young man in the Greek myths who flew on artificial wings to which feathers were affixed with wax. When he flew too near the Sun, the wax melted and he fell into the sea to his death. If anything flies too near the Sun, it is the asteroid Icarus. (Icarus's approach to Earth, however, is no record. Its closest possible approach in its present orbit is 6,400,000 kilometers, or 4,000,000 miles.)

Since 1948 other Apollo-objects have been discovered with perihelia inside Venus's orbit (see Table 50) but of them all,

Figure 35
ORBIT OF ICARUS

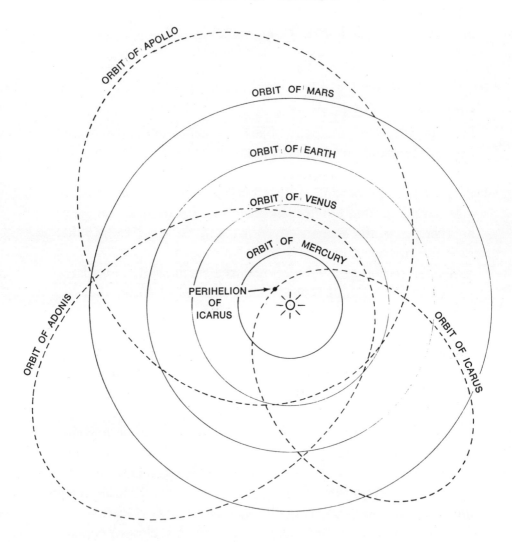

Icarus is still the only one with a perihelion closer than that of Mercury.

A close perihelion does not always mean a close aphelion as

Table 50
APOLLO-OBJECTS

	Perihelion Approach			
	Kilometers	Miles	Venus's Peri- helion = 1	Mercury's Perihelion = 1
Apollo	97,200,000	60,400,000	0.90	2.12
Midas	92,800,000	57,600,000	0.86	2.02
Hermes	86,800,000	53,900,000	0.86	2.02
Cerberus	86,800,000	53,900,000	0.80	1.89
Daedalus	83,800,000	52,100,000	0.77	1.82
1976 UA	70,300,000	43,700,000	0.65	1.53
Ra-Shalom	70,300,000	43,700,000	0.65	1.53
Adonis	65,800,000	40,900,000	0.61	1.43
1974 MA	62,800,000	39,000,000	0.58	1.37
1978 SB	52,400,000	32,500,000	0.48	1.14
Icarus	28,500,000	17,700,000	0.26	0.62

well. The greater the eccentricity of the orbit, the greater the lopsidedness of the two figures. The object 1978 SB, whose perihelion approach is second only to that of Icarus, has an even greater orbital eccentricity than Icarus has, so that its aphelion is farther out than that of any other object in Table 50.

On the other hand, there is Ra-Shalom, discovered in 1978 and named for the Egyptian god of the Sun and the Hebrew word for peace—in honor of the Egyptian-Israeli peace treaty. With a perihelion somewhat farther from the Sun than that of Adonis, Ra-Shalom has an aphelion only a little over a third as far from the Sun as Adonis has. Ra-Shalom and 1976 UA have periods of less than a year. They are the only objects so far known, other than Venus and Mercury, to circle the Sun in less time than the Earth does.

If we imagined ourselves on one of these Apollo-objects, we would have the opportunity to make some unusual observations. Each one crosses the orbit of Earth and Venus, all but two cross the orbit of Mars, and one (Icarus) crosses the orbit of Mercury, while two others skim by it. There are therefore occasions (not in every orbital period, but once in a while) where one planet or another is unusually close and shines brightly in the sky.

Suppose, for instance, there were an observer who happened to be on Hermes in 1937 when it made its close approach to Earth. Earth would seem a starlike object to such an observer most of the time, but on this one pass, Earth would have begun to brighten unusually. It would become more and more brilliant, then begin to show a visible shape as a small crescent.

It would continue to expand, and the crescent would thicken until at closest passage it would be more or less a half-Earth, about 55 minutes of arc in diameter. It would be nearly twice as wide as the half-Moon appears to us and some 40 times as bright. Earth would then shrink as rapidly as it had expanded, while Hermes passed on toward perihelion, not to return to Earth's vicinity for an indefinite number of years.

Each revolution, however, the Sun would expand in Hermes's sky, and in that of any Apollo-object. It is this expanded Sun that would be the most astonishing and impressive object by far (see Table 51).

It would be for Icarus that this effect would be most astonishing and impressive. On Icarus, at aphelion, an observer would see the Sun only half as wide as it appears to us on Earth, and it would then deliver only a quarter as much light and heat as it does to us.

In the course of 204 days, however, Icarus would travel from

Table 51
PERIHELIA OF THE APOLLO-OBJECTS

	At Perihelion	
	Diameter of Sun (minutes of arc)	Light and Heat Received (Earth = 1)
Apollo	49	2.34
Midas	52	2.64
Hermes	55	2.95
Cerberus	55	2.95
Daedalus	57	3.17
1976 UA	68	4.52
Ra-Shalom	68	4.52
Adonis	72	5.06
1974 MA	76	5.64
1978 SB	91	8.08
Icarus	168	27.6

aphelion to perihelion and, in that time, the Sun would swell, slowly at first, but then ever more rapidly as, in approaching the Sun, Icarus's orbital speed became greater. Finally, at perihelion, the Sun would appear over five times as wide in Icarus's sky as it does in our own. Icarus would receive a solar blast more than 27 times as intense as we do, and 2.5 times as intense as even Mercury does at its perihelion (but not quite as great as that which would have been received by the mythical Vulcan).

Icarus rotates in 2.7 hours and has a diameter of about 1.0 kilometer (0.6 mile). This means that if we assume it to be roughly spherical in shape, the surface speed at the equator would be about 1.2 kilometers per hour (0.75 mile an hour).

If we imagined ourselves trapped on Icarus's surface near perihelion, we would be safest if we were on an area where it was just a few minutes before sunrise, because that surface would have had nearly 80 minutes since sunset to cool off. (Such cooling would be rapid in the absence of an atmosphere.)

It would then be necessary, however, to walk westward at the rate of 1.2 kilometers an hour to stay ahead of the Sun. It would be a slow walk, easily managed, but you would have to keep it up for a couple of months without stopping, until the Sun shrank to a size small enough to be relatively safe.

METEORITES

Eleven Apollo-objects are listed in Tables 50 and 51. They range in diameter from an estimated 0.2 kilometer (220 yards) for 1976 UA to 8 kilometers (5 miles) for 1978 SB. More such objects will surely be discovered in the next few years, and many more such objects exist that will not be discovered because they won't happen to come close enough to Earth to become visible.

Some astronomers estimate there are up to 750 Apollo-objects of the size of Icarus, more or less, and there are undoubtedly many thousands that are yards in diameter and perhaps millions that are feet or inches in diameter.

Although the closest approach yet made by an Apollo-object that has been sighted in space was that of Hermes in 1937 (and it is an object about a kilometer across), undoubtedly closer approaches still are made by the more numerous very small Apollo-objects that are too small to see even when they skim by within a few thousand kilometers of the Earth's surface.

Indeed, there are collisions when objects, moving in their orbits, actually cross Earth's orbit at a time when Earth is passing by, and thus enter Earth's atmosphere. For any one object, such a collision is not at all likely; but there are so many small objects whose orbits carry them across Earth's orbit that it is estimated there are some 500 substantial strikes annually.

Once these objects are in the atmosphere, air resistance slows them and their energy of motion is converted to heat so that they glow white-hot. They flash through the air as "shooting stars" and are also called "meteors" from Greek words meaning "things in the air."

Although some of their outer layer is melted and vaporized by their passage through the air, usually a remnant survives and strikes the Earth as a "meteorite." In previous centuries, few such strikes were noted and recovered, but nowadays about 20 per year are recovered and studied. Altogether, fewer than 2000 meteorites the world over have been studied, and of them all perhaps 40 are more than a meter (40 inches) in diameter.

Even an annual bombardment of 500 such small objects is not likely to do damage, since the Earth is exceedingly large and the chance that one of the objects will hit a human being, or even some man-made object, is small. There is no known case of a meteorite's killing a human being, although a woman in Alabama reported being bruised by a glancing blow from one on November 30, 1955. To be sure, as the world's population grows and as the Earth is covered by more and more man-made objects spread more and more thickly, the chances of such strikes increase.

There is also the chance of a particularly large meteorite strike now and then. Large objects are less common than small ones, so the likelihood that a large strike will take place is remote.

Still, Earth's history has been a long one, and there are unmistakable records of large strikes. In Coconino County, Arizona, there is a round crater about 1.3 kilometers (0.8 mile) across and nearly 0.2 kilometer (650 feet) deep, surrounded by a lip of earth about 40 meters (130 feet) high.

Many tons of meteoric iron have been found in its neighborhood, and it seems certain that it is not the result of a volcanic eruption but of a meteorite strike. It is therefore called "Meteor Crater."

It is estimated that Meteor Crater was gouged out by the fall of an iron meteorite about 33.5 meters (110 feet) across, having a mass of about 180,000 tons. The fall took place (at a rough guess) some 25,000 years ago when *Homo sapiens* was on the Earth, but had not yet developed a civilization. No meteorite strike that devastating has taken place since, as far as we know.

Nevertheless other, even larger, strikes had taken place earlier, and if the craters all remained as they were formed, the land surface of Earth might seem as pockmarked now as the Moon or Mercury. The action of wind, water, and life on Earth have, however, erased almost all those craters, although faint traces of some that may be a million years old or more have been detected from the air. Chubb Crater in northern Quebec is over 3 kilometers (2 miles) in diameter and Ashanti Crater in Ghana, Africa, is 10 kilometers (6 miles) in diameter.

Yet, even these are relatively small. What would happen if one of the visible Apollo-objects were to hit us? Hermes and Icarus are 1 kilometer across, and even if they are rock rather than iron, they would be well over 12,000 times as massive as the meteorite that gouged out Meteor Crater. Such a collision could well destroy an American state or a European nation.

Its indirect effects might be worse.

A suggestion has recently been made that a major meteorite strike may have taken place about 70,000,000 years ago. It was sufficiently massive so that its collision with the land threw up a pall of dust that filled the upper atmosphere for three years. The dust cut down the amount of sunlight reaching

Earth to such a degree that plant life failed and life on Earth was brought almost to an end. Most of the species on Earth were indeed wiped out, including the dinosaurs. If such a meteorite had fallen in the ocean rather than on land, it would have splashed up a wall of water that might well have drowned a major portion of Earth's land life.

To be sure, none of the Apollo-objects we have sighted has an orbit that can possibly carry it onto a collision course with Earth, but those orbits are constantly shifting because of planetary pulls. The major axis twists steadily (as that of Mercury does in the effect that gave rise to the thought that Vulcan existed), and since each object sometimes approaches one of the planets fairly closely, additional changes are introduced.

It is estimated that allowing for changes in orbit, there would be a collision with Earth on the part of a sizable Apollo-object, say a tenth of a kilometer in diameter, every 250,000 years on the average.

It is estimated, in fact, that every million years four sizable Apollo-objects strike Earth, three strike Venus, one strikes either Mercury, Mars, or the Moon, and seven have their orbits altered in such a way that they leave the solar system altogether. That, however, doesn't mean that the number of Apollo-objects diminishes with time, for it is also likely that new ones are constantly being added to the list at a rate that has kept the total number constant over the last few billion years.

One possible source of renewal is the asteroid belt (though, as we shall see, it is not the only one). There may be as many as 400,000 objects over a kilometer in diameter in the asteroid belt, and there may occasionally be collisions that send fragments flying in all directions—and some in such a direction as to make them into Apollo-objects.

And as long as we're speculating, we might also suppose that as humanity learns how to make space its home, it will learn how to track down the dangerous objects in space that are uncomfortably large, or come uncomfortably close, and pulverize them while they are still in space by nuclear bombs or by other more advanced means.

9

Comets

THE "HAIRY STARS"

There is another class of astronomical object which, on occasion, approaches the Sun more closely than Venus does.

These are softly shining, hazy objects that stretch across the sky like fuzzy stars with long tails or streaming hair. Indeed, the ancient Greeks called them *aster kometes* (meaning "hairy stars") and we still call them "comets" as a result.

People used to be frightened of comets. They did not follow regular paths across the sky, as the stars and planets did. They came and went unpredictably. The streaming "hair" made some people think of a wailing woman, or else they thought of it as a sword, or as some other unpleasant object. As a result, people were sure that the sudden appearance of a comet meant that some disaster was on its way.

They were right, too, for there was always some disaster when a comet appeared in the sky. On the other hand, there was always some disaster when a comet did not appear in the sky, too, but somehow people didn't seem to notice that very much.

The ancient Greek philosopher Aristotle was of the opinion

that comets were not planetary objects, but phenomena formed within the atmosphere—little glowing patches of air, like will-o'-the-wisps formed miles high. His reasoning was that the heavenly bodies were unchanging and followed regular paths; and since comets underwent peculiar changes and came and went as they pleased, they could not be heavenly bodies.

Aristotle was much respected in later ages in Europe, and for nearly two thousand years his view of comets was accepted. In fact, it wasn't till 1473 that any European did more than shudder when a comet appeared in the sky. In that year a German astronomer, Regiomontanus, observed a comet and put down its position against the stars night after night. That work was the beginning of the modern study of comets.

When a comet appeared in 1532, two astronomers, an Italian named Girolamo Fracastoro and a German named Peter Apian, noticed that wherever the comet was as it crossed the sky, its tail always pointed away from the Sun.

Then, in 1577, another comet appeared, and this time a Danish astronomer, Tycho Brahe, observed it and tried to determine its distance by parallax. If the comet was anywhere in our atmosphere, it would have to be closer than the Moon and its parallax would have to be larger than that of the Moon.

Not so! Tycho found that the comet's parallax was too small for him to measure (telescopes had not yet been invented), so it had to be far more distant than the Moon. That proved that Aristotle was wrong. At least one comet (and, therefore, probably all comets) was outside the atmosphere and beyond the Moon and would have to be accepted as an astronomical object.

That did not end the puzzles concerning comets, for they did not seem to follow paths similar to those of other heavenly bodies. Kepler, who had shown in 1609 that planets went about the Sun in elliptical orbits, didn't think that comets moved in the same fashion. He suggested that they weren't part of the solar system. He thought they might come from far outside the solar system, that they passed through the planetary orbits in a straight line, and that they then vanished into the far distance again.

In 1687 the English scientist Isaac Newton worked out the theory of gravity. Once that was accepted, it was seen that a comet couldn't pass through the solar system in a straight line. The Sun's gravity had to pull at comets and force them to travel in a curved path.

HALLEY'S COMET

Then, in 1705, the English astronomer Edmund Halley studied what records he could find concerning various comets and found that the comets of 1456, 1531, 1607, and 1682 all seemed to have followed the same path across the sky. He noted, too, that they had come at 75- or 76-year intervals.

It struck him that comets circled the Sun just as planets did, but in orbits that were extremely elongated ellipses. They spent most of their time in the enormously distant aphelion portion of their orbit where they were too distant and too dim to be seen, and then flashed through their perihelion portion in a comparatively short time (see Figure 36). In the perihelion portion they were visible, but without the rest of their orbit for study, their comings and goings seemed capricious.

Halley predicted that the comet of 1682 would return in 1758. He did not live to see it happen, but it did return and was sighted on December 25, 1758. It was a little behind time because Jupiter's gravitational pull had slowed it as it passed. This particular comet has been known as Halley's comet ever since. It returned again in 1832 and 1910, and is slated to return once more in 1986.

Figure 36
A COMETARY ORBIT

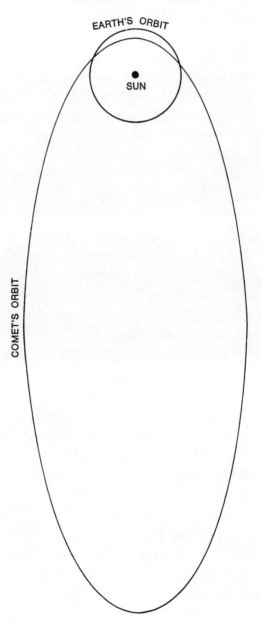

Checking reports of comets back in time, astronomers now believe that Halley's comet has been observed on 32 different occasions since 1058 B.C. Since it has circled the Sun 39 times in that interval, there were seven occasions in ancient times when no one made any record of its arrival—at least none that has survived to this day.

Other comets since then have had their orbits worked out, and comets in general are now accepted as respectable members of the solar system.

A currently popular theory is one advanced by a Dutch astronomer, Jan Hendrik Oort, in 1950. He suggested that in a region stretching outward from the Sun at 7 to 15 trillion kilometers—several thousand times as far from the Sun as even the farthest known planet—are a hundred billion small bodies with diameters which are, for the most part, from 1 to 10 kilometers across. All of them together would have a mass of no more than 1/8 that of Earth.

This material is a kind of "cometary belt" left over from the original cloud of dust and gas that condensed nearly 5 billion years ago to form the Sun and its planets. The outermost fringe of that cloud remained behind in a spherical "shell" about the distant Sun.

The comets differ from the asteroids in that while the asteroids are rocky in nature, the comets are made chiefly of icy materials that would easily evaporate if they were near some source of heat.

Ordinarily, these comets stay in their far-off home, circling slowly about the distant Sun with periods of revolution in the millions of years, and forever cold so that their substance remains icily solid.

Once in a while, however, because of collisions, or because of the gravitational influence of some of the nearer stars, some comets are speeded up in their very slow revolution about the

Sun and, in consequence, move out of the solar system altogether. Other comets are slowed and move toward the Sun, circling it and returning to their original position, then dropping down again. Such comets can be seen when they enter the inner solar system and pass near the Earth.

Because comets originate in a spherical shell, they can come into the inner solar system at any angle. Their orbits can have any inclination to that of Earth and the other planets. Halley's comet has an orbital inclination of 162 degrees and travels in clockwise (retrograde) fashion about the Sun.

Once a comet enters the inner solar system, the heat of the Sun vaporizes the icy materials that compose it, and dust particles trapped in the ice are liberated. The vapor and dust form a kind of hazy atmosphere about the comet, and this is what makes it look like a large fuzzy object.

Thus Halley's comet, when it is completely frozen, is an object only about 2.5 kilometers (1.5 miles) in diameter. When it passes by the Sun, the haze that forms all about it can be as much as 400,000 kilometers (250,000 miles) in diameter, taking up a volume that is over twenty times that of giant Jupiter—but, of course, the matter in the haze is so thinly spread out that the whole thing has hardly any mass.

Issuing from the Sun are tiny particles, smaller than atoms, that speed outward in all directions. This "solar wind" strikes the haze surrounding the comet and forces it outward in a long tail, which can be more voluminous than the Sun itself, but in which matter is even more thinly spread. Naturally, this tail has to point away from the Sun at all times (see Figure 37).

At each pass around the Sun, a comet loses some of its material as it vaporizes and streams out in the tail. Eventually, after a couple of hundred passes, the comet simply breaks up altogether into dust and disappears. This dust, including small bits of gravel, fills the inner solar system and uncounted millions

Figure 37
A COMET'S TAIL

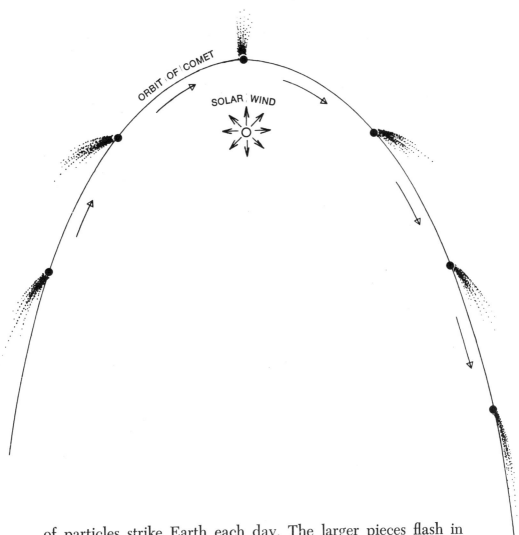

ORBIT OF COMET

SOLAR WIND

of particles strike Earth each day. The larger pieces flash in the sky at night as dim shooting stars, but are completely consumed. None reaches the Earth's surface as anything more than slowly settling dust.

Some comets have rocky cores, and when all the ice is gone the rocky core remains behind and continues to circle the sky as an asteroid. In fact, some astronomers suspect that the various Apollo-objects are comets whose icy portions have long since evaporated, and that it is the comets that serve as the source of new Apollo-objects over the ages.

In the long history of the solar system, many millions of comets have either been speeded up and driven out of the solar system, or have been slowed and made to drop toward the inner solar system where they eventually leave behind their rocky cores or merely a cloud of dust. There are still, however, many billions left. There is no danger of running out of comets.

ENCKE'S COMET

Cometary orbits are of all shapes, because once comets come into the inner solar system they are forever being affected by the gravitational pull of the various planets. Some of them end up with their entire orbits within the planetary system, and they never return to the cometary shell from which they originated.

These are "short-term comets" because their periods are a century or less. Such relatively small orbits are short enough for the detailed orbital figures to be worked out. Of the short-term comets, four, including Halley's comet, at perihelion are closer to the Sun than Venus is, and they therefore should be considered in this book particularly (see Table 52).

With such very large orbital eccentricities, the difference in distance between perihelion and aphelion must be very high and, indeed, that is so (see Table 53).

Halley's comet has the largest orbit of the four. Though at perihelion it is closer to the Sun than Venus is, at aphelion it is

Table 52
SHORT-TERM COMETS

	Period of Revolution (years)	Orbital Eccentricity
Encke's Comet	3.30	0.847
Brorsen's Comet	5.46	0.810
Brorsen-Metcalf's Comet	69.1	0.971
Halley's Comet	76.0	0.967

Table 53
PERIHELIA OF SHORT-TERM COMETS

	Perihelion Distance		Aphelion Distance	
	Kilometers	Miles	Kilometers	Miles
Encke's Comet	50,600,000	31,400,000	612,000,000	380,000,000
Brorsen's Comet	88,200,000	54,800,000	840,000,000	522,000,000
Brorsen-Metcalf's Comet	72,500,000	45,100,000	4,960,000,000	3,080,000,000
Halley's Comet	87,800,000	54,600,000	5,280,000,000	3,280,000,000

farther from the Sun than any known planet but Pluto. (Pluto, at aphelion, is 7,350,000,000 kilometers, or 4,570,000,000 miles, from the Sun, and this is 1.4 times the aphelion distance of Halley's comet.)

At aphelion, an observer on Halley's comet would see the

Sun as a star, but by far the brightest star in the sky. It would shine with a brightness equal to 380 times that of the full Moon as seen from the Earth (though this would still be only 1/1250 as bright as the Sun seen from Earth).

For decades, as Halley's comet moved along its cigar-shaped orbit, the Sun would brighten, but slowly. It would seem scarcely to change from month to month and even from year to year, but careful measurement would show that it was slowly brightening and that the rate of brightening was increasing.

Finally, as Halley's comet moved past the orbit of Jupiter, its speed would be increasing noticeably. It would move past the planets in an increasing hurry, and from its surface the Sun would be seen to grow into a small orb, then a larger one and a still larger one.

By the time Halley's comet reached perihelion, the Sun would have swollen to a width of 54 minutes of arc and would be blasting the comet with 2.85 times the amount of heat that Earth gets. That Halley's comet survives at all is only because it hastens past the inferno very rapidly and then skims out and away from the Sun again. Even so, at each pass it loses a sizable portion of its substance, and as a result it is growing less spectacular.

Encke's comet is, in many ways, the most remarkable of the short-term comets. It was first observed by a French astronomer, Pierre Méchain (may-SHAN) in 1786, but was named for Encke, who studied its motion in detail in 1819.

Encke's comet has the smallest orbit of any comet known and the shortest period of revolution. At aphelion, Encke's comet is not as far from the Sun as Jupiter is, and it is the only known comet with an aphelion inside Jupiter's orbit (see Figure 38).

What's more, passing in the neighborhood of the Sun, as it does every 3.3 years, it has used up almost all its icy substance.

Figure 38
ENCKE'S COMET

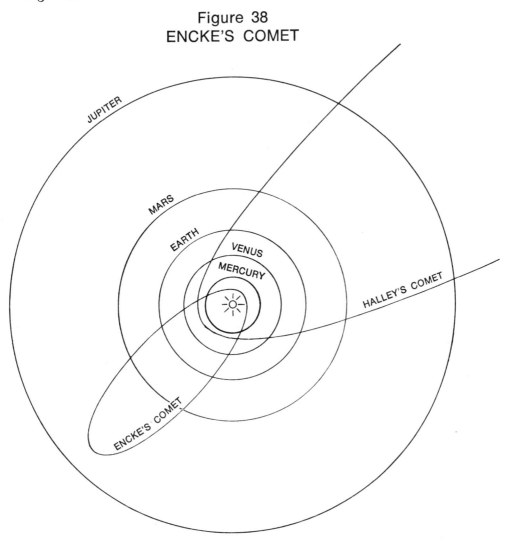

In fact, the only way you can distinguish it from an asteroid is that it is still accompanied by a slight haze. There is no tail, of course; there can be a tail only on relatively icy comets that have not visited the Sun too often. A few more passes and there will be no way of telling Encke's comet from an Apollo-object.

Of all the short-term comets, Encke's comet approaches most closely to the Sun. At perihelion it is closer to the Sun than Mercury is at its aphelion—but it is not as close as Mercury at its perihelion.

The orbit of Encke's comet is very much like that of Apollo-object 1978 SB, and this may not be an accident. When a comet passes perihelion, it can break into two or more pieces. This has actually been seen to happen on several occasions. The different pieces then follow the same orbit. It may be, then, that 1978 SB is a piece of Encke's comet, but one from which all the icy matter has been driven.

It may also be that there are other pieces of Encke's comet following its orbital path, pieces that are too small to see.

This seems probable in consideration of what happened on June 30, 1908, at 7:14 A.M. There was a tremendous explosion in what is almost the exact center of Siberia, in the middle of a vast forest.

All around that explosion, trees were knocked down for a distance of up to 40 kilometers (25 miles). A herd of 1500 reindeer was wiped out. By the most fortunate of chances, however, so desolate was the area that not one single human being was killed.

It was assumed that a large meteorite had struck, or that a volcano had exploded, but no expedition managed to get to the site to study it until 1927. It was then found that there were no signs of any volcano on the site and, what's more, no signs of any meteoric remnants, or any crater gouged out of the ground.

Ever since then, people have been speculating on what might have caused the vast explosion. All sorts of far-out explanations have been suggested, even that of a possible extra-terrestrial spaceship with nuclear-powered engines crash-landing on Earth and setting off a nuclear explosion.

Astronomers reject the far-out explanations. On the basis of the evidence, they reason that a small comet struck the Earth.

A comet, like a meteor, would heat up as it passed through the atmosphere, but the ice of a comet would vaporize much more easily than the rock or iron of a meteor would. In fact, as the comet penetrated denser and denser layers of air and heated further and further, the time would come when all of it would vaporize at once. The heat would produce a fireball, and the expansion would produce a violent thunderclap and shock blast.

There would be most of the effects of a meteor strike, but the comet would explode while it was still in the air and would be even more effective in knocking down distant trees than a meteor strike would be. What's more, after the explosion there would be nothing left: there would be no actual collision with the ground, no crater, no meteorite remnants.

If the data are checked as accurately as possible and if a path is calculated for the object as it approached the Earth, that path seems as though it might be related to the orbit of Encke's comet. There is just the possibility, then, that Earth was struck by a splinter of that object.

SUN-GRAZERS

The short-term comets are not all there are. The short-term comets are, in fact, rather used-up dregs of comets that have been to the well too often. Halley's comet is the only one of them that is still spectacular.

There are, however, comets that are fresh from the vast and distant cloud far beyond the planetary system. There are comets that have not yet had their orbits interfered with by planetary gravitational effects to the point where they have been "captured." They come into the inner solar system, yes, but only at intervals of a million years or more, and spend most of their time in the aphelion portion of their orbits, far beyond Pluto.

When such comets are at their aphelia, the Sun is to them no more than a bright star, no brighter in appearance than Venus at its brightest is to us. The Sun remains such a star without much change for a million years or so, and then it slowly begins to brighten more and more rapidly until it expands in the sky for a brief period, then shrinks rapidly, then more slowly, and becomes nothing more than a star again for a million years or so.

If Earth had such an orbit and if life could somehow be imagined to survive the long drift in outer space and the short, blazing encounter with the Sun, there would be time for human beings to evolve, develop a civilization, and perhaps die out without ever knowing that that bright star in the sky could expand to a blazing inferno (assuming that they didn't develop an advanced astronomy).

These are "long-term comets." Kohoutek's comet, which swooped around the Sun in 1974, was a long-term comet. It was a disappointment because it unexpectedly turned out to be rocky, so it didn't shine brightly and develop an enormous tail.

Many of the long-term comets seem to exist in groups, each a splinter of an original comet that broke up at perihelion. Astronomers recognize fifteen such groups of long-term comets, usually distinguished by letters. Group M make up what are called the "Sun-grazers" because they make an unusually close approach to the Sun at perihelion.

These not only approach the Sun more closely than Mercury at perihelion, but they are also the only known objects that approach more closely than Icarus at perihelion. They approach even more closely than the mythical Vulcan was supposed to. Eight of the Sun-grazers are now known, and the approach at perihelion to the Sun is given in Table 54.

Table 54
THE SUN-GRAZERS

Year of Appearance	Perihelion Distance			
	To Sun's Center		To Sun's Surface	
	Kilometers	Miles	Kilometers	Miles
1668	9,600,000	6,000,000	8,900,000	5,550,000
1887	1,450,000	900,000	755,000	468,000
1965	1,200,000	740,000	505,000	308,000
1882	1,150,000	715,000	455,000	283,000
1945	940,000	585,000	245,000	153,000
1843	820,000	510,000	125,000	78,000
1880	820,000	510,000	125,000	78,000
1963	790,000	492,000	95,000	60,000

The approach is always given to the Sun's center, and even for Icarus the difference between the approach to the center and to the Sun's surface isn't enough to make a fuss over. In the case of the Sun-grazers it is, and in the table both distances are given.

The figures are all but incredible. They make Mercury, and even Icarus, look like objects far distant from the Sun.

When the comet of 1963 was at its perihelion (see Figure 39), the Sun was 122 degrees in diameter, stretching more than two thirds of the way from horizon to horizon and taking up just about half the sky in area. The amount of heat and light it got from the Sun at that time was about 53,000 times as much as Earth gets and about 2,000 times as much as Icarus gets at perihelion.

How do the Sun-grazers withstand the heat? Why don't they just pop and vanish?

Figure 39
COMET OF 1963

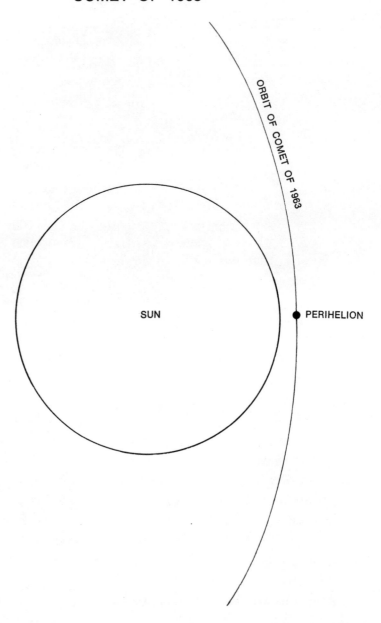

To begin with, they are moving very quickly under the lash of the intense gravity of the nearby Sun. The comet of 1965 was moving at a speed of 100 kilometers per second (62 miles per second) and completed its pass in 6 hours. The comet of 1963 was moving at something like 180 kilometers per second (110 miles per second) and completed its pass in less than 3 1/2 hours.

Second, the cloud of vapor and dust formed by the evaporation of the comet blocks some of the Sun's radiation from reaching what is left of the comet, and helps it survive the relatively short interval when it is nearest the Sun.

Third, the comet *does* suffer severely as a result, and it is doubtful that it can survive more than just a few passes at Sun-grazing perihelion. The original Sun-grazer has broken into eight pieces that we know of, and there may be more. The pieces themselves break up, too. The comet of 1965 broke in two as it passed the Sun. The comet of 1843 vaporized to such an extent that it formed a tail that was visible over a stretch of more than 300,000,000 kilometers (190,000,000 miles), a distance equal to that from the Sun to the asteroid belt.

There we have it, then.

We began with Venus and then asked ourselves whether anything came closer to the Sun than it did.

Within the orbit of Venus there is one planet, Mercury.

Approaching the Sun more closely than Venus over part of their orbits are a few asteroids, of which the most remarkable is Icarus; a few short-term comets, of which the most remarkable is Encke's comet; and a few long-term comets, of which the comet of 1963 seems to hold the all-time record for a close approach to the Sun and survival.

The various asteroids and comets spend only a small portion of their time nearer the Sun than Venus is, however. Most of their time, especially that of the Sun-grazer comets, is spent outside Venus's orbit, and in every case the period of revolution about the Sun is greater than that of Venus—sometimes enormously greater.

Only one object remains inside Venus's orbit at all points; only one object has a shorter period of revolution. That object is Mercury, which holds the record for remaining in the close neighborhood of the Sun *continuously*.

Glossary

ALBEDO—The fraction of sunlight reflected by a planet or satellite.

ANGULAR MEASURE—The measure of length in fractions of a circle: degrees, minutes, and seconds of arc.

APHELION—The point in a planet's orbit where it is farthest from the Earth.

APOGEE—The point in the Moon's orbit where it is farthest from the Earth.

APOLLO-OBJECT—An asteroid whose orbit brings it closer to the Sun than Venus is.

ARGON—An inert gas making up 1 percent of Earth's atmosphere.

ASTEROID—A small planet.

ATMOSPHERE—The layer of gases surrounding a planet, satellite, or star.

AXIAL INCLINATION—The angle between the axis of rotation of an object and a line perpendicular to its plane of revolution.

AXIS OF ROTATION—The imaginary straight line about which an object spins.

CARBON DIOXIDE—A gas found in Earth's atmosphere in small quantities, and in Venus's atmosphere in large quantities.

CELSIUS—A temperature measure in which water freezes at 0° and boils at 100°; 1 Celsius degree equals 1.8 Fahrenheit degrees.

CENTRIFUGAL EFFECT—The tendency for anything spinning about a center to move away from that center.

CIRCUMFERENCE—The total length of the curve of a circle; the length of the largest circle drawn on the surface of a sphere.

CLOCKWISE—Turning in the direction the hands of a clock move.

COMET—An object moving about the Sun in an elongated orbit and made up of "ices" that evaporate and form a tail in the neighborhood of the Sun.

CONJUNCTION—The appearance of a near approach of a planet to the Sun as viewed from the Earth.

CORE—The innermost portion of an astronomical body.

COUNTERCLOCKWISE—Turning in the direction opposite to that in which the hands of a clock move.

CRESCENT—A curved shape like the appearance of the Moon when less than half of it is sunlit.

DEGREE—An angular measure equal to 1/360 the circumference of a circle.

DENSITY—The mass of an object divided by its volume.

DIAMETER—The length of a line passing from one edge to the opposite edge across the widest part and through the center of a circle or sphere.

DIRECT MOTION—Moving counterclockwise.

EARTH-GRAZER—An asteroid that approaches the Earth more closely than the planet Venus ever does.

ECCENTRICITY—The degree to which the ellipse of an orbit is flattened.

ELLIPSE—A curve that looks like a flattened circle.

EQUATOR—The circumference that lies halfway between the two poles of a spinning object.

EQUATORIAL BULGE—The extra thickness of a planet in the equatorial regions due to the centrifugal effect of its rotation.

EQUATORIAL DIAMETER—The diameter stretching from a point on the equator to the opposite point on the equator.

EQUATORIAL SPEED—The speed of motion of a point on the equator of an object as that object rotates.

ESCAPE VELOCITY—The speed at which an object must move to escape from the gravitational pull holding it.

FAHRENHEIT—A temperature measure in which water freezes at 32° and boils at 212°; 1 Fahrenheit degree equals 5/9 of a Celsius degree.

FOCUS (plural, foci)—One of two points inside an ellipse. The two foci are at equal distances from the center of the ellipse and on opposite sides, along the major axis.

FORMALDEHYDE—A gas made up of molecules containing one carbon atom, two hydrogen atoms, and one oxygen atom; once thought to occur in Venus's atmosphere.

GIBBOUS—A shape similar to the Moon's when it is larger than a half-Moon but smaller than a full Moon.

GRAVITATION—The attraction exerted by one object on the other objects in the universe.

GRAVITATIONAL LOCK—A gravitational effect which causes an object to rotate in a simple fraction of the time in which it revolves about another object.

GREENHOUSE EFFECT—A warming effect that follows when light and heat can pass through a barrier into an object, but not out again.

HORIZON—The most distant point on a landscape, where it appears to meet the sky.

HYDROCARBON—A substance made up of molecules containing hydrogen and carbon atoms only.

HYDROGEN—A gas composed of the simplest of all atoms.

INFERIOR CONJUNCTION—The position of a planet when it is closest to Earth, between the Earth and the Sun.

INFRARED RADIATION—Light-like radiation made up of waves longer than those of visible light.

KILOMETER—A measure of length equal to about 5/8 of a mile.

LIBRATION—The back-and-forth movement of an object in the sky resulting from the uneven speed of the body on which the observer is located.

LONG-TERM COMETS—Comets with orbits too elongated for a definite determination to be made of their period of revolution.

MAGNETOSPHERE—A region outside a planet where particles from the solar wind are trapped along that planet's magnetic lines of force.

MAGNITUDE—A figure representing the apparent brightness of an object shining in the sky. The lower the figure, the brighter the object.

MAJOR AXIS—A diameter passing through the foci and center of an ellipse.

MANTLE—The portion of the Earth, Venus, or Mercury that surrounds the core.

MARIA—Large, relatively flat areas of the Moon that originated as lava flows.

MASS—In a general way, the amount of matter in an object.

MAXIMUM ELONGATION—The greatest separation of a planet from the Sun in Earth's sky.

METEOR—The streak of light that occurs when a small bit of matter from space passes through Earth's atmosphere.

METEORITE—A small body from space which has collided with Earth's solid surface.

MICROWAVES—A form of light-like radiation used in radar, with waves longer than those of infrared radiation.

MINOR AXIS—A diameter at right angles to the major axis of an ellipse, and the shortest diameter of the ellipse.

MINUTE OF ARC—An angular measure equal to 1/60 of a degree.

MOLECULE—A group of atoms that are held together in ordinary matter.

NITROGEN—A gas making up 4/5 of Earth's atmosphere and also found in Venus's atmosphere.

NODE—The place where the plane of one orbit crosses the plane of another.

OBLATENESS—The flattening of a planet from spherical form because of the centrifugal force of rotation.

OPPOSITION—The appearance of a planet in the part of the sky opposite to that in which another object such as the Sun appears.

ORBIT—The path taken by an object revolving about a larger object.

ORBITAL INCLINATION—The angle made by the orbital plane of a planet with the orbital plane of Earth.

ORBITAL PLANE—The flat surface that would result if every point on the orbit of a planet about the Sun were connected with the center of the Sun.

ORBITAL SPEEDS—The speed with which a body travels along its orbit.

OXYGEN—An active gas making up 1/5 of the Earth's atmosphere.

PARALLAX—The apparent change of position of a close object compared to a more distant object, when the viewer shifts the position from which he views the object.

PERIGEE—The point in the Moon's orbit where it is nearest the Earth.

PERIHELION—The point in a planet's orbit where it is nearest the Sun.

PERIOD OF REVOLUTION—The time it takes an object to make one complete turn about a larger object.

PERIOD OF ROTATION—The time it takes an object to spin once on its axis.

PHASES—The different shapes of the lighted part of a planet or satellite that is shining by reflected light from the Sun.

PLANET—A body that circles a star and shines only by reflected light.

PLATE TECTONICS—The very slow motions of the various sections of the Earth's crust, and the effects of those motions.

POLAR DIAMETER—The diameter stretching from the north pole to the south pole.

POLES—The two points where the axis of rotation reaches the surface of a rotating body.

POTASSIUM—A common metal, one variety of which slowly breaks down to argon.

PROBE—A rocket-driven vessel designed to pass near some planet or satellite in order to gather information about it.

PROGRADE MOTION—Moving counterclockwise.

RADAR—A device making use of microwaves to determine the direction and distance of objects capable of reflecting those microwaves.

RADIO WAVES—Like microwaves, but longer.

RETROGRADE MOTION—Moving clockwise.

REVOLUTION—The circling of an object about another object.

ROCKET—A jet engine that is thrust forward by the discharge backward of hot gases produced by combustion.

ROTATION—The spinning of an object about its own central axis.

SATELLITE—An object that revolves about a planet.

SECOND OF ARC—An angular measure equal to 1/60 of a minute of arc.

SHORT-TERM COMETS—Comets whose orbits lie within the planetary system and with a period of revolution of a hundred years or less.

SIDEREAL DAY—The length of a planet's period of rotation relative to the stars.

SOLAR DAY—The length of a planet's period of rotation relative to the Sun.

SOLAR SYSTEM—The Sun and all the objects that are held in its gravitational field and that move around it.

SOLAR WIND—Charged particles from the Sun moving out at high velocity in every direction.

SPECTROSCOPY—The technique of dividing light from any object into its component wavelengths.

SPECTROSCOPY, INFRARED—The analysis of the infrared light that arrives from the planets to find out which wavelengths are present or missing.

SPECTRUM (plural SPECTRA)—Light that has been spread out so that each different wavelength is in a different position.

STAR—A mass of matter much larger than a planet, within which nuclear fusion takes place causing it to grow hot and glow.

SUBATOMIC PARTICLES—Particles smaller than atoms.

SULFUR—An element that is a yellow solid at earthly temperatures.

SULFUR DIOXIDE—A gas made up of molecules of a sulfur atom and two oxygen atoms; occurring in Venus's atmosphere.

SULFURIC ACID—A corrosive liquid with molecules made up of two hydrogen atoms, a sulfur atom, and four oxygen atoms; found in Venus's cloud layer.

SUN-GRAZER—A comet which, at perihelion, comes closer to the Sun than Icarus does.

SUPERIOR CONJUNCTION—The position of a planet when it is on the directly opposite side of the Sun from Earth.

SURFACE—The outside of any solid object.

SURFACE GRAVITY—The strength of the gravitational pull on the surface of an astronomical body.

SYNODIC PERIOD—The time from one conjunction of a planet to the next conjunction.

TELESCOPE—A tube, containing lenses or mirrors, that makes distant objects look larger, nearer, or brighter.

TIDAL EFFECT—The differential pull of a body on various parts of another body circling it that may slow the rotation, round the orbit, or alter the distance of that body—or all three.

TRANSIT—The movement of an object between the Earth and the Sun across the face of the Sun.

ULTRAVIOLET RADIATION—Light-like radiation with waves shorter than those of visible light.

VOLUME—The room taken up by any object.

ZENITH—The point in the sky that is directly overhead.

Index

About the Author

ISAAC ASIMOV is a versatile and prolific writer, the author of more than 225 books on subjects ranging from science fact and fiction to history, humor, and literature. *Venus, Near Neighbor of the Sun* is his fifth Lothrop astronomy book; previous titles are *Alpha Centauri, the Nearest Star; Jupiter, the Largest Planet; Mars, the Red Planet;* and *Saturn and Beyond.*

Born in Russia, Isaac Asimov came to the United States at the age of three and grew up in Brooklyn, New York. At Columbia University he earned the degrees of B.S., M.A., and Ph.D., and for nine years taught biochemistry at the Boston University School of Medicine, where he still holds the title of professor.

A noted wit and raconteur, he is in great demand as a speaker, and has received countless awards and honors. He is a scientist, a science writer, a futurist, and one of the great science fiction writers of all time. He and his wife live in New York City.